SLICK WATER

Andrew Nikiforuk

SLICK

WATER

Fracking and One Insider's
Stand Against the World's
Most Powerful Industry

DAVID SUZUKI INSTITUTE

 GREYSTONE BOOKS

Vancouver/Berkeley

Doreen Docherty—my profile in courage.

Greystone Books Ltd.
www.greystonebooks.com

David Suzuki Institute
219–2211 West 4th Avenue
Vancouver, BC, Canada V6K 4S2

Cataloging data available from Library and Archives Canada
ISBN 978-1-77164-076-3 (cloth)
ISBN 978-1-77164-077-0 (epub)

Editing by Barbara Pulling
Jacket and text design by Nayeli Jimenez
Jacket photograph by iStockphoto.com
Printed and bound in Canada by Friesens
Distributed in the U.S. by Publishers Group West

Canada

We gratefully acknowledge the financial support of the Canada Council for the Arts, the British Columbia Arts Council, the Province of British Columbia through the Book Publishing Tax Credit, and the Government of Canada through the Canada Book Fund for our publishing activities.

Greystone Books is committed to reducing the consumption of old-growth forests in the books it publishes. This book is one step toward that goal.

"*Slick Water* is a true-life noir filled with corruption, incompetence, and, ultimately, courage. It is a deeply informative, disturbing, and important book."

ELIZABETH KOLBERT, author of *The Sixth Extinction*

"Andrew Nikiforuk crafts a stunning picture of fossil fuel industry and government abuse, telling the story of fracking through the lens of an unprecedented legal battle to expose the truth. As this engrossing book shows, the more we understand that the culture of extraction thrives on the worst kind of exploitation and violence against all of our communities, the broader and deeper the sisterhood of resistance becomes."

NAOMI KLEIN, author of *This Changes Everything* and *The Shock Doctrine*

"In this balanced and meticulously researched page-turner, Andrew Nikiforuk follows Alberta whistleblower Jessica Ernst's quest for justice with a passion, verve, and level of damning detail reminiscent of A Civil Action."

JOHN VAILLANT, author of *The Jaguar's Children*

"Nikiforuk masterfully reveals what everybody in the oil and gas industry knows, but does not want the public to know, about past, and likely repeat, offenses."

ANTHONY R. INGRAFFEA, Dwight C. Baum Professor of Engineering Emeritus, Cornell University

"Through the eyes of one intrepid activist, Nikiforuk reveals the dark side of the oil and gas industry's latest 'success' stories."

RICHARD HEINBERG, Senior Fellow, Post Carbon Institute

"The arc of the moral universe is long, but it bends
toward justice."

MARTIN LUTHER KING, JR.

"The Orwell law of the future: any new technology
that can be tried will be. Like Adam Smith's invisible hand
(leading capitalist economies toward ever-increasing wealth),
Orwell's Law is an empirical fact of life."

DAVID GELERNTER,
"The Second Coming: A Manifesto," *Edge*, 1999

"Thousands have lived without love, not one
without water."

W. H. AUDEN, "First Things First"

CONTENTS

PROLOGUE

A**T THE BEGINNING** of the American Civil War, the humorist
Mark Twain trekked to Nevada, where he tried his hand at
mining. Like most fortune seekers, Twain hoped to
become a nabob living "flush times." But Twain lost money
and failed miserably at getting rich. Shortly afterward, the
writer coined a new definition for mining: "a hole in the
ground with a liar on top."

Scholars now debate whether Twain made the remark.
Some suggest the quote may have originated with a man from
Kansas. But the definition stuck and has been used ever since,
because it remains an apt description of mining.

This book is an unconventional true story about a new
form of disruptive mining: hydraulic fracturing.

ONE

The Dress
for Less Explosion

O N THE AFTERNOON of March 24, 1985, the Ross Dress for Less clothing store at West 3rd and Ogden in the Fairfax District of Los Angeles blew up into an inferno. A spark from somewhere—most likely a janitor flicking a basement switch—ignited an invasive cloud of methane with a hellish boom. The explosion ripped up concrete slabs, blew out windows, singed cars, and cracked walls. "Fire belching fissures" opened in the ground. The conflagration injured twenty-three shoppers, some of whom suffered third-degree burns. It also forced the evacuation of twenty to thirty stores along a quarter-mile commercial strip. One witness said the sky was "raining fire." Another compared the disaster to an earthquake. "Out of nowhere the room exploded," a shopper told a TV news crew, "and I was thrown to one side." For days, escaping plumes of methane gases licked the foundations of buildings and sidewalk cracks. Firefighters tried in vain to extinguish the flames.

The explosion did more than rattle the residents of Los Angeles—it inaugurated the continent's first debate about hydraulic fracking, horizontal drilling, and migrating gases. The oil and gas industry blamed the explosion on bacteria and naturally seeping gases. Scientists, however, reckoned differently. They identified hydraulic fracturing and abandoned leaky oil wells as the chief suspects.

Like any good detective tale, the Dress for Less story begins in an earlier time: California's black gold rush. In the early 1900s, wooden oil derricks popped up so fast in Los Angeles that streets looked like a surreal forest painted by Picasso. Some two hundred oil companies drilled more than two thousand wells into scores of different oil fields. The frenzy turned the city into a "vibrant oil-soaked little canyon." Early homeowners gamely hosted wooden derricks in their backyards and disposed of oil-field waste in their basements. Oil wells planted in cemetery plots allowed the dead to provide income for the living, in the form of royalty checks. The thirty-year-long boom manufactured big fortunes for Union Oil, Getty Oil, and Atlantic Richfield. The flowing oil dollars in turn fueled the state's Wild West car culture and enriched evangelical churches. Petrodollars supported the upstart film industry and a raft of corrupt real estate deals. Oil turned California into an early American Kuwait.

The oil frenzy produced some memorable characters, just as the shale gas boom would do one hundred years later. A religious piano teacher named Emma Summers became "California's Oil Queen" thanks to her "genius for affairs." She made millions selling oil to local hotels and industry. A dairy farmer, Arthur F. Gilmore, discovered black gold in the Salt Lake Oil Field in 1902 while looking for water. The field, which occupied an area about one mile by two miles, soon became the highest-producing formation in California. By 1917, more than four hundred wells had extracted an astounding 50 million barrels of oil from the highly faulted field, where tar seeped to the surface at Rancho La Brea. Gilmore got rich from his find and started his own oil company, advertising, "Someday you will own a horseless carriage. Our gasoline will run it." Gilmore Oil later pioneered the original "gas-a-teria," the first pump-your-own gas station.

3

After Gilmore's oil wells started to go dry in the 1930s, a farmer's market and a stadium arose on top of the Salt Lake Oil Field and its abandoned oil wells, sump pits, and oil spills. The city's restless economy planted another crop of buildings on the site as well, including the famous CBS Television City and a shopping center. No one dreamed of the possible consequences, because an oil-fueled city rarely sleeps.

Nor did the oil industry abandon the city's subterranean oil fields. During the 1960s, many companies reentered the Salt Lake field with horizontal wells or long wells drilled on a slant. To this day, industry extracts 28 million barrels of crude a year from LA's petroleum basement. Land owned by the Archdiocese of Los Angeles sports scores of oil and gas wells, as do several movie lots. Industry disguised the Cardiff pumping station as a synagogue. In a 2010 exhibit on "urban crude," LA's Center for Land Use Interpretation revealed that thousands of wells puncture the city "like tree roots," extracting "the living essence of the ground, fueling this city of the car." Moreover, the exhibit added, industry drains the "progenerative substrate" by operating "in cracks, corners, and edges, hidden behind fences, and camouflaged into architecture, pulling oil out from under our feet."

Three days after the Dress for Less explosion, the city struck a task force to determine the cause. The investigating body included representatives from Los Angeles's Department of Building and Safety as well as the Division of Oil and Gas (the state's energy regulator) and the Southern California Gas Company. Task force investigators consulted experts in geophysics and performed a chemical analysis on the methane that had set the block ablaze. Meanwhile, an emergency crew drilled an eighty-foot-long well under the parking lot of the Ross Building, where the store had been. They found a pocket of highly pressurized gas, which they vented and flared off.

Three months later, the task force issued a lengthy report that blamed Mother Nature. The report explained that a "pressured incursion of naturally-occurring, almost-pure methane gas" had seeped through small openings between the floor slab and the foundation into the store. The probable source "was not from an oil well but from decomposing organic matter nearer the surface." The report suggested the methane gas was "formed from the decomposition of buried plant materials at no deeper than 100 to 200 feet below ground level." Higher-than-average rainfall had raised the groundwater, it explained. Elevated water levels had pushed pockets of methane made by bacteria into the "path of least resistance," which just happened to be the store. Furthermore, this random movement of gas to the surface had been taking place for tens of thousands of years and would, "in all likelihood, continue to occur indefinitely into the future." To reduce the risk of another incident, the city needed to take steps "to help prevent gas from entering into a building or, if the gas does enter a building, detecting it before an explosive gas level is reached." The report recommended "the venting of significant paved parking areas," too. In response, throughout the Fairfax District, the city installed basement gas detectors and a variety of pipes to vent methane.

Incredibly, the task force said little about the most likely 5 suspect in the explosion. Even the Southern California Gas Company had initially admitted that "the flames were fueled by gas from a long-abandoned oil field in the vicinity." But the task force regarded the fact that the Salt Lake field lay under the store as incidental geography. The report considered it an aside that approximately 500 abandoned oil and gas wells punctured the neighborhood over 1,200 acres, and that another forty active wells still pumped oil, water, and gas out of the ground. The abandoned wells, which lay 6 to 10 feet below buildings, yards, and streets, had been improperly plugged, with redwood fence posts or debris. Most, if not all, leaked methane. But they weren't the issue, assured the report. Their existence was simply an unhappy coincidence.

A year after the Task Force on the March 24, 1985 Methane Gas Explosion and Fire in the Fairfax Area exonerated the oil and gas industry, its members quietly tabled two commissioned yet contradictory methane fingerprint studies. One, by GeoScience Analytical, took fingerprints of gas from eight different areas in Los Angeles with a history of seepage and abandoned wellbores. Although the samples all came from hilly locations, the study said most of the methane appeared to have originated in swamps. It supported the task force's blame-it-on-bacteria conclusion. But the other study, which the task force never published, offered a radically different conclusion. That study looked at the chemical makeup of gases found licking sidewalks outside the Dress for Less store. Those fingerprints didn't look anything like swamp gas. But they did match deep gas from the oil reservoir being pumped nearby. The samples also contained ethane and propane, clear markers of gas originating in a deep oil deposit.

6   Nobody saw the second study except the lawyers and technical consultants representing citizens in a class action lawsuit. Shortly after the Dress for Less explosion made national news, one of the badly burned casualties, a CBS employee, contacted Matthew Biren, a personal injury lawyer. Biren didn't know anything about the oil and gas business, but he recognized a clear case of negligence. He signed up twenty-one of the shoppers hurt by the blast and eventually launched two class action suits: one against the city of Los Angeles and, later, one against McFarland Energy.

Biren warned the victims, as he instructed all of his clients, that life wasn't fair, and that such lawsuits rarely succeeded. Nonetheless, he recruited an incredible team of technical experts. They included Richard Meehan, a Stanford University expert in fluid migration, and Douglas Hamilton, a prominent civil engineer, along with the University of Southern California's George Chilingarian, one of the world's most famous petroleum geologists. Chilingarian, a polymath who started university at the age of twelve, had located many of the world's most productive petroleum oil fields, and he edited a variety of key oil journals. His father had served as physician to the shah of Iran. "Chilingarian had written about forty books and knew everybody in the business," Biren later recalled. With Chilingarian's contacts, Biren recruited a gas storage and migration expert from Paris, as well as Denis Coleman and Martin Schoell, U.S. geochemists who were pioneers in gas fingerprinting. Coleman had just formed Isotech Inc., a firm that specialized in tracking down different sources of methane.

Biren also hired Bernard Endres, a brilliant systems safety engineer and lawyer.

Meehan, an experienced industry troubleshooter, knew a thing or two about the idiosyncratic behavior of aging oil fields, and none of the facts for the Ross Building added up. Meehan had sharpened his skills while exploring the causes of the Baldwin Hills disaster twenty years earlier, with colleague Douglas Hamilton. In 1963, a new $10 million dam burst asunder in Los Angeles. The colossal rupture sent 292 million gallons of water spilling into a residential community, destroying hundreds of homes and killing five people. Lawsuits pinned the blame for the disaster on the oil and gas activity just south of the dam. Industry silenced that talk with an out-of-court settlement. However, Hamilton and Meehan later proved that fluid injection, combined with sinking ground, had played a major role in fracturing the dam.

Their research broke new scientific ground. During the 1950s, Standard Oil had begun injecting millions of barrels of brine water at high pressure into the Inglewood Oil Field to coax out more petroleum. Due to excessive pressures, the firm's injection wells often leaked or blasted through to other producing oil wells. (Today industry calls the costly debacle "interwell communication.") Five of nine injection wells located near a natural fault close to the dam had lost their fluid just prior to the rupture, and brine waters had burst up through the ground like mysterious fountains. Standard Oil's massive water pumping had eliminated the shear strength along preexisting faults and had fractured the formation, creating "new fracture surfaces." The researchers concluded that "although fluid injection operations may be carried out for beneficial purposes, the effects of such injection on the geologic fabric can be serious and far-reaching." Their paper, published in *Science* in 1971, was one of the first to warn that hydraulic fracturing came with unanticipated geological consequences.

8   As Matthew Biren's team got to work, more methane burped to the surface in LA's Fairfax District. This time a pedestrian spotted a small geyser of gas, water, and mud at the southwest corner of the Gilmore Bank building. Investigating firefighters discovered near-explosive levels of methane in the bank, a school, and a Kmart store. Once again the city evacuated and cordoned off the district. And once again the city struck a task force to "protect health and safety within the area."

The second task force discovered that the gas-venting wells installed in 1985 had already clogged with sand. The blockage had forced methane to migrate through soils and into nearby buildings. Because there was no rising groundwater table this go-round, the second task force reported that the "origin of methane gas is complex and not clearly understood." However, the report said, it really didn't matter whether the methane came from swamps or an aging oil field, because the gas behaved in a flammable manner either way. The second report mentioned abandoned oil and gas wells as a possible source of the methane, but it concluded that "all available chemical and isotopic information" exonerated the aging infrastructure. The task force noted that forty-two active wells were still pumping salty wastewater into formations underneath the neighborhood to help bolster oil production, but that wasn't a problem either; these operations did "not appear to have an adverse effect on the area methane seepage." One regulatory member of the task force chastised the group in a letter accompanying the report for daring to mention the region's oil legacy.

By contrast, a letter written by James E. Slosson, a former state geologist, criticized the task force for disassociating methane seepage from oil and gas activity in the Salt Lake field.

To Richard Meehan and Douglas Hamilton, both of the city's blue-ribbon reports seemed wedded "to a heavily edited script with a happily blameless ending." For starters, the swamp-gas story wasn't accurate. Dennis Coleman's tests on samples of methane taken from Dress for Less in 1985 and 1989 conclusively showed that the methane contained ethane and most certainly originated in an oil and gas field. "There is no evidence of near surface microbial origin," declared Coleman. After reviewing regulatory data on well production data in the district, Meehan and Hamilton learned that McFarland Energy had drilled several horizontal wells running miles in every direction from an enclosure surrounded by eucalyptus trees. The camouflaged drilling island sat right next to the Fairfax farmer's market. The firm's aging wells pumped nearly a thousand barrels of salty water for every barrel of oil extracted, and as a consequence, McFarland had a growing saltwater disposal problem. Forbidden by the city to dump the brine water into storm sewers, the company had transformed an old oil well into an injection site. In January 1985, it fractured the well with hydrochloric acid with the goal of cracking open more space in the rock to hold all the unwanted saltwater.

But the new disposal well, Meehan and Hamilton discovered, had changed the pressure in the oil field, which in turn affected natural faults, some of which ran underneath local streets. In a 1992 paper published in *Engineering Geology Practice in Southern California*, the scientists pieced together the final story. Gas had migrated to the surface via "passages temporarily opened along the Third Street fault through the process of hydraulic fracturing." In the case of the Dress for Less explosion, the 3rd Street fault had "fractured enough to allow upwardly migrating pulses of gas-pressurized water and free gas" to enter the soil or build up under the pavement. "It seems to us," Meehan and Hamilton concluded, "that future studies of the Los Angeles gas problem can hardly be considered complete without full investigation of these issues."

George Chilingarian, who would later write a bestselling book on gas migration, agreed. The gas that blew up the department store had strayed from oil formations along a combination of man-made and natural fractures, he said, including a leaky wellbore. The official explanation that methane originated from plants was preposterous. The world-famous petroleum hunter told Biren he was prepared to testify in court that if the methane gas that caused the explosion came from organic matter, then "we've solved the energy problems of the world. All we have to do is plant a few plants."

But the information amassed by Biren's experts never saw a judge or a jury. In 1990 McFarland Energy, Inc. settled out of court. The company didn't want any more public attention directed to the horizontal wells that spidered underneath the city in all directions. McFarland Energy paid the twenty-one injured citizens—all of whom signed confidentiality agreements—somewhere between $2.5 and $8 million but admitted no liability. Russell Wollman, an attorney representing the firm, assured the LA Times that the company's experts, if given the chance, would have proved that the methane came from decomposing plants.

Matthew Biren had conclusive scientific data proving otherwise. Speaking to the LA Times, he welcomed the settlement as "an appropriate acknowledgement on the part of McFarland of its responsibility." But he worried that an enduring methane hazard remained in the district. Lawrence Hirst, vice president of McFarland Energy, dismissed the notion and repeated that methane seeps were normal. By cleaning up a naturally occur- 11 ring nuisance, Hirst added, "we'd suggest that as a producer of oil and gas in the area, we are actually helping to alleviate the problem."

The following year, Bernard Endres and George Chilingarian published a paper entitled "Environmental Hazards of Urban Oilfield Operations" in the *Journal of Petroleum Science and Engineering*. Using the Dress for Less explosion as one example, they explained that "many oilfields located in urban settings are managed by catastrophe rather than through preventive management." In contrast to industry claims that drilling reduced the risk of gas seepage, the scientists wrote, the practice actually did the opposite. As oil wells aged, the pressure in the reservoir dropped, and gas that had initially comingled with the oil separated out and became "available for migrations." The stray gas often wandered up poorly sealed wellbores, they said, due to lousy cement jobs. "Although corrosion is recognized in the oil industry as the number one problem in causing deterioration of well casing, companies frequently operate using a catastrophe management approach, not taking action until a serious leak has already occurred." Injecting wastewater into older oil wells, they found, exacerbated gas migration by creating more fractures. Their conclusion was unequivocal: "Clearly, there is a great need for a uniform set of procedures and guidelines to be established for the monitoring of dangerous levels of gas seepage and land subsidence, especially in urban areas where the surface dwellers usually have no idea of the hazard that underlies them.

They are generally helpless to take action, even if they become aware of the hazard."

Endres and Chilingarian would go on to repeat their warnings in later scientific papers. Their research also debunked the industry claim that most of the methane contaminating groundwater had diffused through rocks and popped to the surface naturally over time. Such a process did indeed take place, they found, but not on a human time frame. The researchers estimated that it might take methane trapped in a reservoir 1,740 meters (5,709 feet) deep about 140 million years to move upward, and even then it would travel only in small quantities. Methane sought out fractures or faults "as pathways for migration," however, and in many cases fractures created by oil-field activity or man-made channels such as wellbores did the job by serving as convenient "pipelines" for not only methane but also radon and carbon dioxide (CO_2). The draining of oil fields caused the ground to subside, and that too created fractures and pathways for gas migration. According to Endres and Chilingarian, the only way to protect the public was to mandate extensive testing for soil, groundwater, and air toxins, as well as the monitoring of oil and gas well leaks. "Oil and gas wells must be carefully evaluated," they warned. "Old wells must be reabandoned to protect against the risk of oilfield gases migrating up the old wellbores and entering the near surface environment."

In 1993, Martin Schoell and Dennis Coleman summed up their findings on the Dress for Less explosion in a publication of the Society of Petroleum Engineers. Gas from swamps possesses a light and dry fingerprint, the methane trackers wrote. In the Fairfax District, the heavy, wet fingerprint of methane collected from seeps and venting gas wells not only precluded "a bacterial origin" but directly matched "local oil-field production."

Hamilton and Meehan's 1992 paper remains the most critical summary of the incident. Attributing the cause of the event to the "digestive rumblings of an ancient and invisible swamp" or portraying it as "yet another production of Los Angeles' quirky environment," they wrote, conveniently absolved the world's most powerful industry of any liability. The task force's conclusion that the leaking methane was natural not only eliminated "troublesome legal issues." But it also implied "that the methane hazard could exist virtually anywhere, so no human agency was at fault for its workings."

JESSICA ERNST HAS read hundreds of scientific papers on the industrial migrations of methane including the paper by Hamilton and Meehan. The oil patch consultant knew methane leaks from oil and gas fields were not only a chronic hazard, but a multibillion dollar liability. In 2007 she filed a $33 million lawsuit against the oil giant Encana and two provincial regulators, Alberta Environment and the Energy Resources Conservation Board, for gross negligence. Her legal ordeal began after Encana intentionally fracked community aquifers near her home in central Alberta and contaminated the fresh water with explosive levels of methane and industry-made toxins. What she thought was an alarming case of groundwater contamination turned into an epic struggle against the world's most powerful industry and its enablers.

Ernst had read about the Dress for Less explosion disaster years earlier, but it wasn't until 2012 that she connected the dots. As she reread the studies on the LA disaster, Ernst now zeroed in on the troubling similarities to her own case.

13

Meehan and Hamilton's reference to "a heavily edited script with a happily blameless ending" particularly rang through her head. Industry and regulators had known about the dangers of gas migration from oil and gas fields since the 1980s, but they had forged a legal script for avoiding all liability: blame it on bacteria. For decades corporations had pretended that nothing could be more natural than the discovery of explosive levels of methane in water wells and buildings. Alberta authorities had used exactly the same argument to dismiss Ernst's and other well-water contamination cases. The Dress for Less tale, realized Ernst, now underscored the true nature of her transgression: she had amassed a damning volume of scientific evidence that discredited industry's claims. She almost felt like ripping her skin off, and she hadn't felt that way for a long time.

14

TWO

This Much
You Should Know

IN A MOMENT of weakness, and she does not have many, Jessica Ernst will confess, "From time to time, even though I know without a doubt in my heart that I am meant for this lawsuit, and that I have no choice, I must do it, my mind and fears try to sway me away. My ego fights with my heart, exclaiming: Why should I have to give up everything for others? 'Cause by the time this lawsuit is done, my life will be, too, or near to done."

And then Ernst will go for a walk along a prairie river, where she will sometimes find an animal skull lying on the ground. Ernst likes bones, because they remind her of our fleeting connection to all things living and dead. Bones, she says, can quell her anxiety.

"And then I remember money is nothing. My belongings are nothing. Water is everything. Truth is much more important." In these renewed moments of strength, Ernst no longer frets about spending her life savings to compel her government to enforce its own laws. She needs little from life now, she says, and wants for even less.

There are a few more things you need to know about Jessica Ernst.

She is tall and big-boned, with a head of striking white hair. She arrived with flaming red hair in Montreal General Hospital on May 15, 1957. White-blonde hair quickly replaced the red mop. She is the first "legitimately born" female child in her family for four generations. Ernst loves her younger brother, Martin, but she does not talk to him. Like many Albertans, Ernst was once married. She says she cherished the man, but she never liked the idea of marriage. She hasn't been romantically involved with anyone since 1999.

Ernst has lived most of her adult life in the oil-exporting province of Alberta, which has been ruled by one political party for forty-three years. She compares the province's dysfunctional political culture to that of a Soviet republic. Sooner or later, everything gets compromised by the revenue generated by hydrocarbons. She describes Alberta as a crazy, greedy, and infuriating place. When she was married, Ernst lived on a beautiful acreage in the Alberta foothills. She later moved to a condo in the city of Calgary, Canada's ode to Houston. Now she inhabits a farmstead at the edge of badlands, where dinosaur bones and coal seams poke out of outcrops. That's where her water got fracked, sending her life on an unexpected course. Her small bungalow faces west and looks up a coulee etched by wind and sun. Her property sits on the outskirts of the pretty hamlet of Rosebud. She bought the old farm for its solitude and its water but laments, "Now that they've fucked my water, my anonymity, privacy, and solitude are gone too."

Friends of forty years, and she has many, describe Ernst as precise, blunt, determined, damaged, charming, organized, generous, demanding, exacting, practical, forthright, and meticulous. She keeps records on everything from the mushroom species in Rosebud to the restoration of well sites in northern British Columbia. She has a memory of steel (friends call it a "rattrap") and an unnerving eye for detail. Those who know her well say that her greatest burden is also her most powerful strength: she forgets nothing. Ernst can be an exasperating and unrelenting critic for either friend or foe. Evidence and truth trump any relationship. A perfectionist, she freely admits she does not suffer fools gladly, and she makes no apologies for being rude to assholes.

With good reason, she is wary of middle-aged German men and males in crowded rooms. She admits that she really trusts no one. She believes that if you want something done, you had better do it yourself.

Ernst can outcurse any oil-rig worker, and her irreverent sense of humor often startles friends. Even her protracted legal case is fair game for her jests. She enjoys sarcasm and at times sparkles with mischief. But she also possesses a Germanic sense of right and wrong. The words *justice* and *fairness* pepper her vocabulary, though she is no do-gooder: "I'm a Taurus," she explains, "and very, very stubborn." Ernst wears baggy clothes and does not like to socialize much. Her mother considered her to be mildly autistic. As a child Ernst would hide from visitors by retreating into a cupboard, and as an adult she would rather mine a mountain of data to establish an obscure fact than mingle with a room full of people. She works to the point of deathly exhaustion. She rarely takes the easy road. The lawsuit she has launched against the world's most powerful industry does not surprise her friends. They concur that government and industry have unwisely underestimated the woman: "She makes a formidable enemy."

As a child, Ernst passed the marshmallow test, which means she is an expert at delaying gratification. In the 1960s, Stanford University psychologists tested the will of children by placing subjects in a room with a marshmallow for fifteen minutes. The children were told that if they didn't eat the marshmallow, scientists would reward them with another one. Two-thirds of the children gulped the marshmallow in seconds. Only a few had the willpower to wait. "Jess could have sat there for four hours. She was always like that," says Helen Rezanowich, who grew up with Ernst in Montreal's west island. "She was the kid who had Halloween candy six weeks after the fact." The Stanford psychologists tracked their subjects into adulthood, and they found that the children who could delay gratification demonstrated greater success as adults at navigating life's problems. At the time of writing, Ernst's lawsuit has lasted seven years. Her lawyers warn it may go on for another seven. "I have a roomful of marshmallows," says Ernst.

Ernst says the greatest thing she has ever done was to quit smoking, in 1989. The second greatest thing was to quit drinking, in 1993. The third was to file a $33 million lawsuit against the natural gas giant Encana, Alberta Environment, and Alberta's Energy Resources Conservation Board for gross negligence. In US terms, that's the equivalent of a Texas landowner suing Exxon Mobil, the Railroad Commission, and the Government of Texas.

JESSICA ERNST'S FAMILY history is complex and sorrowful. She is the daughter of immigrants whose lives were fractured by Hitler's Germany. What generations of coal mining had not already undermined in her family, the war perforated. Her mother, Gerda, who was part Jewish, grew up in Langenbochum, a coal-mining center. During the war a wealthy Dutch family hid the young Gerda from the Nazis in their home.

Ernst remembers visiting her great-grandmother's dark tenement in Langenbochum in the 1960s during two trips to Europe. The apartment complex housed workers who toiled at the Hammer and Chisel mine. Industry had excavated so much coal underneath the town that the land around it slouched like a broken man. Some of the apartment buildings tilted 45 degrees. Most buildings had no bathtubs or toilets. The poor have always raised their own cheap meat, and the tenement's residents stored their washtubs in the complex's rabbit hutches. Whenever someone wanted a bath, they hauled the tub upstairs. Ernst recalls that the buildings reeked of urine, alcohol, and coal. Like her tenement relatives, Ernst now hauls water by truck to take a bath. She also composts waste from her toilet in the garden. Her well water is just too explosive.

Ernst's father, Ludwig, or "Lutz," grew up in Westerholt, another dark mining town. Lutz came from a middle-class German family. He initially studied to be a priest but couldn't stomach the hypocrisy. "Everyone must find their own way to God," he once told Jessica. "Never listen to church or a religion. Find your own way." Lutz became a bricklayer instead. During the war he fought as a German paratrooper but was captured by the French. His captors offered him a choice: starve or join the French Foreign Legion. In Indochina he became a prisoner again. After the war he married Gerda. The two survivors fled the ruins of fascism and moved to Montreal.

Everyone in Ernst's family seems to have walked a tragic road. Her maternal grandmother, a German Protestant, fell in love with a Jewish lawyer from Holland before World War II. The man killed himself in grief because they could not marry due to the religious conventions of the day. He died not knowing that his sweetheart was pregnant with Jessica Ernst's mother, Gerda.

Gerda had two children herself, but loved only one: Jessica was the unwanted child. Her mother said the girl was ugly and plotted to send her away from the age of eight, often farming her out to other families for the summer. As a consequence, Ernst spent much of her childhood with her Oma and the flinty Polish man her grandmother had married before immigrating to Canada. They lived nearby on the West Island of Montreal. Ernst was happiest climbing trees or swimming in the wild currents of the Rivière des Prairies, which divided the island of Montreal, and in which six of her neighbors accidently drowned. She would steal bouquets of lilacs from abandoned houses and bring them to her grandmother.

Her Oma, a gentle but sad woman, often read stories to Ernst, including "Little Red Riding Hood." Oma loved the tale, but Ernst hated it. She preferred *Winnie the Pooh*. Her Oma also taught her to recite a traditional bedtime prayer: "*Ich bin klein* (I am small), / *Mein Herz ist rein* (My heart is pure), / *Daß niemand drin wohnen als Jesus allein* (Nobody may dwell in here but Jesus)." At the age of nine Ernst stopped reciting the prayer. Something stole away her childish exuberance, and she became serious and shy. Today, before giving a talk or boarding a crowded plane, Ernst will recite another one of Oma's sayings. It sounds eloquent in German but reads awkwardly in English: "*Immer wenn man denkt es geht nicht mehr, kommt von irgendwo ein LichtLein her.* (Every time you think you can't possibly go on, comes from somewhere a tiny light.)" In 1982 her

Oma could no longer find the light. She walked into the Rivière des Prairies in the middle of the night and drowned.

Ernst's mother, Gerda, vivacious and outgoing, worked as a
hospital cleaner in Montreal. She later sold perfume at Dorval
Airport. Gerda learned English by reading *The Merchant of Venice*,
and she named Jessica after Shylock's only daughter. Jessica's
father, Lutz, worked far from home in an iron ore mine in Lab-
rador. He later built the family a house in the west end by a big
weeping willow tree and tried his hand at running a gas station. Whenever Lutz calculated that the station might finally
turn a profit, BP asked him to put up a new sign or install a new
gadget.

As the debts mounted, Gerda and Lutz's marriage eroded.
Lutz would go strange after one bottle of beer, and he drank
constantly. Ernst remembers a lot of yelling in the house. As
Gerda grew louder, Lutz retreated into a silence bordering on
invisibility. There were infidelities. The two divorced in 1968,
when Jessica was eleven years old. One day Gerda drove Ernst
and her brother, Martin, over to their grandmother's house,
explaining that they had to leave their dad, their house, and
their beloved guard dog, Fipse, behind forever. Although Lutz
wrote regularly, Gerda destroyed the correspondence. She also
burned his album of war photographs. "Gerda didn't want to
live with a man who went bankrupt; she wanted success in her
life," says Ernst.

After the divorce, Ernst lived largely with her Oma for most
of her teenage years and worked at a market garden to help
support the household. For 75 cents an hour, she hoed vegetables and tended cattle, sheep, and ducks. As a young woman,
Ernst swore to herself that she would never be beholden to
any bank, company, or entity that could take away a home or a

business. She became a studious saver. Freedom from debt, she says, gave her the freedom to pursue the lawsuit. She does not know many people who are free.

Jessica's mother taught her that the job of womenfolk was to cook, clean, and take care of children. When Jessica announced to Gerda one day that she was planning to go to university, Gerda said, "No, you won't." Jessica proved her mother wrong, studying environmental biology at the University of Guelph. Fascinated by the transmission of disease and parasites, she majored in entomology. She later studied the transmission of heartworm disease in dogs and graduated with a Masters in Science from the Ontario Veterinary College.

In 1982 Ernst drove west in a white Toyota two-wheel-drive truck to find work. Alberta didn't need any bug experts at the time, but the oil patch wanted land agents to do seismic work. Ernst signed on as a field supervisor and permit agent for Summit Land Consultants. In short order, she learned how to broker a deal with angry landowners when oil and gas companies doing seismic work destroyed fences, killed livestock, or sullied water wells. The point, she says, was to be honest and keep your word. On one job a rig hand, fresh out of jail, punched her in the mouth. Ernst was undaunted. In 1987 she and her new husband, Sean, started their own land agent business: O'Neill and Ernst Permitting.

The business flourished, but the marriage failed, and in 1994 Ernst started her own firm, Ernst Environmental Services. She specialized in assessing the impacts of pipelines and wells on fish, valleys, rivers, and archaeology sites. One of her jobs was to restore to wilderness a road that had been built by Imperial Oil to a now-abandoned well site. She eventually reseeded parts of it by hand. Through her dealings with trappers, First

Nations, rural communities, and bureaucrats at the National Energy Board, she established a reputation for getting pipeline applications done on time.

Oil and gas companies made a lot of mistakes in those days, and Ernst dutifully cleaned up many of them. She made sure that roads and fences were repaired, that land was reclaimed, that livestock was replaced, and that landowners were properly paid for their leases. When Ernst started in the business, she recalls, companies such as PanCanadian, a regular client, readily admitted to their errors and hired people to set things right with landowners. "Now," she says, "the industry is in total denial about everything."

23

Denial takes many forms, as Ernst knows all too well. She collects art, or at least she did. She has twenty pieces by Marianna Gartner, a great contemporary painter. All are slated to be sold to fund the lawsuit. Gartner, born in Winnipeg, takes old photos and uses them to create otherworldly portraits of "disturbing ordinariness." Some of her recent paintings display skeletal remains and organic matter together. For Ernst, the skeletons "remind us how empty our bodies really are and what is important is our spirit." A painting called *Skull Girl* covers most of Ernst's living-room wall. It shows a young girl in a white dress skipping a rope on the lone prairie. A human skull covers the girl's face. Ernst bought the painting the same year Encana fracked local aquifers. She believes that in *Skull Girl*, Gartner unwittingly captured the state of Ernst herself as a nine-year-old. Ernst says that's the year she died inside.

And here's another thing you should know about Ernst: she believes that children who are raped by men never heal. They simply learn how to survive and, if God permits, how to forgive those who violated them. Ernst will tell you that time does not heal all wounds and that the proverb is fucking bullshit. Time may soften the pain, but it does not erase the memories. Time may separate the past from the present, but it does not mend what has been broken. Time may heal physical wounds, but it does not repair the violation of trust. Only solitude and quiet and birdsong can heal such pain, says Ernst. That's what her land at Rosebud offered her: asylum from the wreckage.

Memories of Ernst's abuse as a child crashed her life in 1991. A wave of shame and anger overwhelmed her. She spent more than a year in a psychiatric ward in the Holy Cross Hospital in Calgary, battling anxiety and flashbacks.

When she was six, a school bus driver molested Ernst. She was the last kid on the bus route. Two years later a former war buddy of her father's, a man named Rex, raped her while her father lay passed out on the couch. Months later Gerda sent a protesting Ernst to live with the violator's family in Mississauga, Ontario, over the summer. It was her mother's complicity, says Ernst, that murdered the girl inside her. The pedophile's repeated molestations inverted the nine-year-old's womb and damaged her pelvis and neck. (To this day, Ernst's gait remains awkward, and she lives with chronic pain.) During the rapes Ernst stared at the ceiling tiles and counted the dots. For months afterward, she says, "I couldn't live with my own skin. I wanted it off." The man's wife said nothing about the abuse. She called the police only later, when she caught Rex in bed with a nine-year-old boy.

More rapes followed. Abused children give off the scent of wounded animals, Ernst says, and that's the worst thing about pedophiles: they will hunt you down. Ernst still finds nothing more terrifying than the sound of a man's footsteps thumping down the hall at night. The rapes explain why she doesn't trust anybody, she says, and "why I can't stand touring and traveling. The confinement." She still struggles with hugs or any kind of touching.

With years of therapy, Ernst methodically rebuilt her life. The business she started, Ernst Environmental Services, won contract after contract. Ernst was often the only woman working in remote oil and gas camps staffed by hundreds of men. (Her psychiatrist called it "implosion therapy.") In 1994 she appeared on the cover of the Oil & Gas Journal after working on environmental approvals for Home Oil's Kahntah plant and pipeline project in northern BC. For that remote development, she advised Home Oil executives, "If you want to complete the job on time, you had better tell First Nations what you are doing and tell them the truth." Some company representatives objected to sharing their plans with "fucking Indians," but Ernst overruled the racists. Her company started to specialize in pipeline applications and cumulative impacts on wildlife.

Today, Ernst suspects that industry and government have illegally obtained copies of her hospital records. She believes that someday they will use the contents to try to discredit her and the lawsuit. "But I'm not afraid," she says. "I believe that the men who raped me are living a life of hell. I am not."

Her doctor, a kindly man, told Ernst that her recovery would take time. Most rapes are planned acts of violence, engineered to control and degrade, he explained. It was not enough to strip away the anger and shame; she would have to rebuild her heart's ability to assert and defend herself. But the hardest part was this: Ernst could not understand why her parents had failed to protect her.

When she was in her early thirties, Ernst searched for and found her father. Lutz had remarried a younger woman, and his wife felt threatened by Jessica's presence. "I told her that I wanted, needed, nothing from him but to know him," Ernst says, but the woman did not believe her. Nor could her father comprehend the full scale of the horrors visited upon Ernst as a child. A reunion that had begun joyfully ended with Lutz's new wife evicting Ernst from the house. Lutz watched and said nothing. They never saw each other again.

Ernst's psychiatrist told her there would come a day when she realized that the horrors of her childhood had given her a strength few people have, and that she would find something of value in the abuse. For years the statement infuriated and rankled her. But now she recognizes its truth. "The main reason I can do this is that there is nothing anyone can do to me that hasn't already been done. They raped our aquifer. The enablers said you can't talk about it, but Encana broke all the laws and rules. I couldn't stop the bus driver or Rex, but I can do something now."

Jessica Ernst knows the oil patch inside out. For nearly thirty years, she worked for some of North America's most prominent energy companies, including Encana, Esso, Statoil, Murphy Oil, and Chevron. With equal aplomb, she told executives to smarten up and rig hands to fuck off. Her company earned a reputation for promptness, efficiency, and accuracy. Ernst had a habit of telling dark suits in skyscraper boardrooms what the uncomfortable science said about the impact of their activities on wildlife, water, and land. For the life of her, Ernst swears, she does not know how to tell people what they want to hear. Thanks to the lawsuit and the punitive nature of the Alberta government, no company has called on her firm's services since 2011. Old clients told Ernst the word had gone out that no one was to employ her company. She has lived on her savings since then.

Many people in the oil and gas industry consider Ernst's 27 lawsuit a genuine threat to the powers that be. Some also call her "crazy" or a "flake." Energy In Depth, an oil and gas propaganda website, wrote in 2012, "No one is going to rely upon her as a credible source, at least no one who doesn't claim to have been abducted by aliens." Tony Allwright, an Irish petroleum engineer, typically dismissed Ernst as a "professional protester" whose data-filled talks earned her "a reasonable living and an exciting international jet-set life-style." (Allwright got it all wrong: community groups organized her unpaid talks and in many cases did not cover all of her travel expenses.) The Alberta government, a jurisdiction corrupted by petroleum dollars, has accused Ernst of humiliating its investigators and of threatening the very economics of hydraulic fracking. A senior provincial deputy minister once phoned Ernst during a speaking tour Jessica was on and told her to shut up. (Ernst kept on talking.) Encana, Ernst's former client, claims that any problems or injuries suffered by Ernst as outlined in the lawsuit are the result of a "pre-existing medical, psychological or other condition." Ernst's lawsuit makes no mention of health matters.

Ernst has grown accustomed to government and industry officials libeling and slandering her. In the middle of the night, anonymous phone callers have warned her, "Drop your lawsuit or bad things will happen." Her beloved dog, Bandit, was thrown under a train. But the worse things got, the calmer Ernst became. "It is an incredible achievement in life to find and keep contentment," she says, "while watching what one created, achieved, saved, all vanish."

Every time Ernst's case goes to court, landowners as far away as Ireland, Texas, and Michigan pray for her. She has encouraged rural audiences in New Brunswick and New York with an ancient proverb: "Many fleas make big dog move." Her defense of groundwater has unwittingly made her a folk hero on three continents. Poems have been written about her courage and stamina. An Alberta cowboy said that Ernst needs a wheelbarrow to carry her balls, because they are slightly "larger than Superman's." Others describe her as a Valkyrie, one of the female Nordic spirits who rode flying horses into battle. A friend once sent Ernst a postcard of an Emily Carr painting portraying the ogress Dzunuḵwa, a singing spirit of the woods with snakes in her hair. The card read: "I am Ernst. Do Not Fuck with Me." And that was *before* the lawsuit.

Ernst doesn't go to church, but she believes there are consequences for what we choose to do and not do in this life. Ministers, priests, and nuns have called her lawsuit a stirring example of community-mindedness. Ernst finds inspiration in Hildegard of Bingen, the Benedictine abbess and mystic who wrote music, poems, morality plays, and botanical texts. Even popes feared the tiny genius. "The earth is... mother of all that is natural, mother of all that is human," Saint Hildegard believed. She warned that men should not destroy the mothers who sustained them.

Audiences describe Jessica Ernst's inspirational presentations as dense and technical; they often grasshopper from one point to another. In Balcombe, England, wealthy stockbrokers, lawyers, and homeowners listened to the scientist in stunned silence. A blogger later reported, "One of her more telling comments was, 'When you listen to the [industry] promises, and later get the data, they are not the same.'" In Ireland, the media refers to Ernst as a "world-famous activist." (Ernst says she is no such thing.) When she gave a talk called "Life Inside a Frac Experiment" in County Leitrim, Ireland, in 2012, the local paper described her as an "enemy of the state." In Balcombe and New Brunswick, violent battles between protesters and police erupted over fracking just months after Ernst presented talks there. You should know that Ernst abhors violence and has never advocated violence in any form.

Every morning and evening, Ernst drinks tea made from nettles that she harvests on her own land. Nettles, she says, are a girl's best friend. She picks and dries horse mushrooms (*Agaricus arvensis*), which she cooks in butter and garlic. She describes herself as both a hunter and a gatherer. Chocolate and apples remain her favorite foods. Ernst often refers to herself in the third person. She speaks four languages—English, German, Dutch, and French—and when stressed sometimes gets them mixed up. Her favorite time to walk is twilight, when the prairie sky glows in Halloween colors. She believes the best way to face trouble is to confront it directly. Her favorite First Nations saying is "Follow the raven into the shadow and you will find the light."

A cedar waxwing dining on Saskatoon berries can enchant Jessica Ernst. She says the song of the goldfinch is her favorite music. In the early years of dealing with flashbacks from her childhood rapes, she listened to Beethoven's Piano Concerto No. 5 in E-flat Major several times a day on a small cassette player. She considers the 1961 recording by Van Cliburn to be the best. The "Chorus of the Hebrew Slaves" from Verdi's opera *Nabucco* can render Ernst motionless because, she says, it's about never giving up. And whenever the weight of her $33 million lawsuit catches her breath, she listens to the acoustic version of Emmylou Harris's "Home Sweet Home": "And it's a long hard road we walk / On our way to home sweet home."

30   Some nights she plays the song forty times in a row.

THREE

Fracking Oildorado

FRACKING HAD A bloody birth just a decade after the first U.S. oil boom, in the late 1850s in Pennsylvania. So many men and women flocked to the western part of the state to get rich that lawless towns such as Pithole and Oil City sprang up overnight like mushrooms. According to local reporters, the greasy boom towns often smelled like "a camp of soldiers with diarrhea."

The new industry turned farmland into industrial lots crowded with wooden derricks and dotted with abandoned holes. Teamsters commandeered rivers and roads with skiffs and wagons laden with oil barrels. Massive petroleum spills polluted ponds, rivers, and creeks. Men with "oil on the brain" died in fiery explosions. With the help of transient workers, faraway speculators rewrote the state's economy in much the same way that the Marcellus shale gas boom would do 140 years later. The brute nature of mining "rock oil" stunned nineteenth-century farmers. They gaped at the devastation and searched for words: "The earth seems to bleed like a mad ox, wrathfully and violently." A newspaper editor compared the sound of oil drilling to "a huge giant seized with the pains of death" whose agony threw "all nature into convulsions."

Pennsylvania's Oildorado made the United States the world's top oil supplier for several decades. It injected petroleum into the bloodstream of American life, paving a national "road to comfortable affluence." After the Civil War, even religious leaders predicted that petroleum would become America's greatest resource, due to its "recuperative energy." Oil provided cheap illumination for kerosene lamps and lubrication for all manner of machines. Chemists seemed to discover another use for rock oil every day, including in soaps and paints.

Boosters claimed there was enough petroleum in the ground to "fill all the lamps the universe could manufacture and to grease every axle on this revolving planet." But *Scientific American* (then a weekly journal of practical information) had its doubts. In 1866, the magazine lamented that few people had any idea of petroleum's growing importance. Since 1862, oil sales had quadrupled to $77 million a year, and the magazine's editors worried about demand outstripping supply. They wondered how long the oil would continue to flow in Pennsylvania.

The key obstacle appeared to be anemic oil wells. Some went dry quickly, while others got blocked by "extraneous substances" such as paraffin wax. Industry seemed to tap more "dusters" than it did gushers. After a decade of extraction, so many abandoned derricks littered rural Pennsylvania that visitors described them as "decaying monuments to small fortunes ruined." But *Scientific American* offered a marvelous solution: the Roberts Petroleum Torpedo. The technology, explained the magazine, "would fracture the rock, and clear the closed passages or make artificial ones reaching the oil veins."

The idea was not new. Individual drillers in Pennsylvania and New York had thrown jars and canisters of explosives into oil and water wells for years, with unpredictable results. Some explosions made the oil flow, but others damaged the entire enterprise. The future presidential assassin John Wilkes Booth, who owned several oil wells, tried to detonate an explosive on a disappointing oil property in Franklin, Pennsylvania, in the early 1860s. The shot ruined the well.

Colonel Edward Roberts, a former dental technician, arrived propitiously on the scene to turn well-shooting into a standardized monopoly. After being cashiered from the Union Army for drunkenness, Roberts ventured to Oildorado with his brother in 1864. The colonel had brought along a couple of torpedoes, and he quickly smelled an opportunity. The Civil War veteran claimed he first got the idea while watching cannonballs shatter canals during the battle of Fredericksburg, Virginia. The water dampened the concussion, forcing the energy out in a butterfly-like formation that cracked the canal's stone. Roberts decided to test the idea on a water-filled oil well, hoping to "open fissures where the oil comes through." The process required the careful lowering of an iron or tin cask filled with gunpowder two or three hundred feet into the well. A weight called a "Go Devil," connected by a string to the percussion cap, was dropped on the torpedo to set off the explosion. Misfires with torpedoes would subsequently fill graves and became the stuff of legend.

Roberts tried his first "shooting" in 1865 at the Ladies Well and then promptly patented the process, which included the use of a torpedo, a propellant (black powder, and later nitro), and water. The results energized the infant petroleum industry. Torpedoing turned cranky oil wells that had yielded only three barrels a day into sensations gushing as many as eighty. Producers clamored for more torpedoes the way the industry would demand "hydrafracs" in the 1950s or slick-water fracks from horizontal wells in the 2000s. "Next to the discovery of oil, no invention has done more to enrich well owners, than the Roberts Torpedo," proclaimed the Titusville *Morning Herald*.

The fracking of early oil wells attracted large crowds. Some spectacular events used as many as one hundred cans of nitroglycerin. An explosion typically sent up a column of water eight or ten feet high, followed by a mighty roar as trapped gases escaped. Then, wrote one observer, "a column of oil, rushing swifter than any torrent and straight as a mountain pine, united derrick floor and top." When the shot took effect, the derrick looked like it had been "smitten by the rod of Moses." Drillers sometimes dammed rivers and creeks to prevent the splurting oil from washing away. Tens of thousands of wells were torpedoed in Pennsylvania alone.

But Roberts's monopoly on the technology created as much controversy in northwestern Pennsylvania as modern multistage hydraulic fracking would 150 years later. Newspaper reporters dubbed it "the torpedo war." The Roberts Petroleum Torpedo Company not only charged a fee of $100 to $200 per detonation, but demanded one-fifteenth of the proceeds from the increased flow. (The cost of modern fracture treatments ranges from $10,000 to $6 million.) Speculators and drillers balked at the price and were soon hiring "moonlighters" to shoot their wells in the dark of night. The moonlighters slung two 10-quart cans of nitroglycerine in gunny sacks on their backs and devised their own triggering methods. Many didn't live long and instead were "scattered to the four winds of heaven." Their work lit up the night sky in the state's oil fields like some otherworldly battlefield.

To shut down the moonlighters, Roberts hired the Pinker- ton National Detective Agency, along with an army of spies. "You could not spit in the street or near a well after dark" without hitting one of Robert's emissaries, reported a journalist of the day. Roberts got injunctions against competitors and tied his rivals up in the courts until they settled. Employing lawyers from Philadelphia, Pittsburgh, and New York, he threatened as many as two thousand prosecutions and collected millions in damages. By one account, Roberts "was responsible for more lawsuits than any other man in the United States." (The oil industry pioneered, if not established, America's fondness for litigation.) Before the colonel died in 1881, the torpedo man had spent a quarter of a million dollars on lawsuits. Two years later, the U.S. Congress refused to renew his torpedo patent on the grounds that it had caused too much strife. Shortly afterward, a hundred companies entered the trade, with names like The Independent Shooters and The Young Torpedo Company. DuPont created the American Glycerin Company to keep the companies supplied with explosives. "So the darkest chapter in petroleum history, a flood of litigation, a mass of deception, a black wave of treachery and a red streak of human blood, must be charged to the account of Nitro-Glycerine," concluded the oil driller and reporter John McLaurin in 1896.

Well-shooting continued to dispatch its nervy practioners with gruesome efficiency. With nitroglycerine, the trade also introduced a new menace to the countryside. Farmers avoided wagons carrying nitro and kept their distance from the industry's explosive caches and magazines. In the 1870s, the well-shooter Royal "Doc" Wright, along with a telegraph agent, visited a magazine in the woods to retrieve some nitro. While swinging a hatchet to loosen a can of glycerine from a block of ice, Wright missed. As McLaurin reported, "Wright's gold watch, flattened and twisted, was fished out of the Allegheny, two-hundred yards down the stream, in May." The remains of the two men barely filled a cigar box.

THERE WERE A lot of cigar-box funerals in Pennsylvania thanks to torpedo fracking. One shooter left a can of nitro in the bush to retrieve the next day. A family of berry pickers found the can and, thinking it was lard oil, brought it home. The father, George Fetterman, used one drop to grease the axel of an engine in rapid motion and was noisily ushered into the hereafter with a "head crushed into jelly." In 1871, A. S. West, an agent of the Roberts Petroleum Torpedo Company, set off to shoot one last well. The vexing industry retired him first. According to the Titusville Morning Herald, the torpedo "was being drawn up again by Mr. West, and had nearly reached the top, when it exploded, fracturing his skull, breaking some of his limbs, and otherwise injuring him. His death was instantaneous."

William H. Payne, another hapless agent, was blown to bits near Rouseville a year later while transporting cans of nitro and a torpedo to a well site. His singed horses miraculously survived the blast. But Payne's body decorated the forest. The Titusville Morning Herald dutifully reported what searchers had found in the woods: "A portion of the face, including nose, mouth, beard, and one eye, to which was held by a shred a portion of the shoulder blade; the right hand badly mutilated; a portion of the bowels; a few broken ribs; a small piece of the spinal column; his feet without the toes; and the boot with some of the toes in."

The paper later editorialized about the public hazards of well-shooting: "The men who make it their business to fire torpedoes and handle nitroglycerine take their lives in their hands, and are supposed to be prepared for the consequences, but when we consider the large amount of this subtle element which is daily carried through our public thoroughfares and over our public highways it is high time that some stringent measures should be adopted to mark the wagons which carry it, so that all persons having a regard for their personal safety may keep out of its reach."

37

One of the most famous well-shooters was Charlie Stalnaker. The Montana man offered his services in North America's western oil patch on both sides of the border, stimulating exhausted wells back to life. The debonair shooter drove a red car with the word *nitro-gylcerine* emblazoned on it; he was rarely tailgated. When not shooting wells, he'd blow up rattlesnake dens or unplug sewers. To fracture a formation, Stalnaker used anywhere from a few quarts of nitro "soup" to several thousand, depending on the depth of the well. Every once in a while he blew out the windows of nearby homes or sent well casings shooting like rockets into the sky. Fifteen men trained with "Nitro Charlie," but only Stalnaker himself died of natural causes.

All of this activity, however, couldn't overcome certain stubborn geological realities in North America. Although torpedoes extended the life of some oil wells, the brute technology couldn't mask the reality of domestic oil depletion, any more than anti-aging pills can stop time. The flow of oil or gas from a field follows a pimple-shaped progression. Once production peaks, it declines at a predictable rate, no matter how many torpedoes the industry throws down a well. As new oil booms erupted in Texas, Oklahoma, and California, other states wrestled with exhausted formations. Oil production in Pennsylvania topped out in 1896. (In 2013, the state's industry pumped only a sixth of the oil it had in 1891.) Oil production peaked in Ohio in 1886. In New York, it petered out in 1937.

The industry tried various other aids to extend the life of wells. In the 1890s in Lima, Ohio, frackers experimented with acids. A chemist working with the Standard Oil Company tried pouring pressurized hydrochloric acid into limestone formations (60 percent of the world's oil reserves lie in carbonate rock) so that "long channels could be formed." The acid eroded, etched, and fractured the rock. But handling corrosive chemicals in the field proved tricky. The acids often set off unwanted chemical reactions and plugged wells. In the 1930s, Pure Oil and Dow Chemical tried again, this time in Midland, Michigan. Using arsenic as a corrosion inhibitor, company chemists shot down five hundred gallons of hydrochloric acid into a dead well. The new corrosive torpedo did its job. The acid shock revived the well and turned it into a sixteen-barrel-a-day producer. Soon thousands of wells were undergoing "production enhancement with acid stimulation." Dow Chemical and an entity called the Halliburton Oil Well Cementing Company championed the treatment.

But fracturing with acids remained a hit-or-miss prospect due to mechanical problems, inadequate pump pressure, and what industry still calls "the wrong acid package." "Wormholes" started to pop up; injected streams of acid selectively ate away natural fractures in the rock, often creating channels up to twenty feet long that swallowed the rest of the package.

Rust from pipes sometimes changed the nature of chemical reactions and suffocated wells with goop. Even contemporary industry textbooks admit that, with acid fracking, "both success and failure are common." 39

The development of modern hydraulic fracturing, another hit-and-miss technology, took decades. Serendipity, combined with the growing scarcity of U.S. domestic oil fields, played a major role. By the 1930s, U.S. oil companies had run right out of oil gushers. Firms like Standard Oil moved to Venezuela and the Middle East, where they found lots of cheap oil. In Texas and California, anxious producers fretted about their sputtering wells. They turned to fracking for relief the same way Pennsylvania firms had invested in the torpedo.

The initial art was entirely accidental. Sometime around World War I, petroleum drillers observed something intriguing. When they injected water into oil wells, the formation often absorbed more water than they had expected. Researchers suspected that excessive water pressure broke more rock open down the hole, where the fluids leaked off. Later field tests, by Floyd Farris at the Stanolind Oil and Gas Company in Tulsa, Oklahoma, proved the point. Farris solved another puzzle, too. For years drillers had squeezed cement into wellbores to seal them off from aquifers and lock the pipe into place. But some wells ate cement like crazy and required repeated pourings at great cost.

To figure out what was going on, Farris drilled a ten-foot well into sandstone then injected a batch of cement. In digging up the well, he discovered that the pressure of the flowing cement had split open the sandstone, with cracks running in different directions. In one case, the cement had traveled five feet through the rock like splattered pancake batter.

Farris didn't stop there. He wondered what might happen to the rock if he pumped down a different fluid—for example, napalm-thickened gasoline mixed with sand. Napalm, a chemical leftover from the war, was cheap and abundant, and it mixed well with water. It also slid down a wellbore with ease. Farris thought the added sand might prop open the fractures created by the injection of the toxic fluids. The first experiment, in a west Kansas gas field, pumped down a thousand gallons of napalm, two thousand gallons of gasoline, and a batch of sand from the Arkansas River. The frack job failed, and no gas flowed.

Undeterred, Farris tried again, with a "hydrafrac" job in East Texas. This time he combined crude oil, soap, sand, and solvents. The treatment turned a nonproducing well into a solid fifty-barrel-a-day wonder. Farris filed his patent in 1948 and awarded the Halliburton Oil Well Cementing Company a proprietary license to frack like mad. Subsequent tests showed that hydraulic fracturing could boost production by as much as 75 percent. Fracking was taken up in a big way in Pennsylvania, home of the torpedo method. Industry used oil and kerosene to fracture rocks in gas formations in the western part of the state. Some treatments boosted production substantially.

Over the next two decades, industry enthusiastically fracked 500,000 largely vertical and shallow wells with all sorts of fluids, including water, napalm, crude oil, kerosene, diesel oil, fuel oil, and even blackstrap molasses. A 1953 Halliburton advertisement in the *Oil and Gas Journal* vowed that its HydraFrac treatments could "split any formation" and "boost productive flow." To get the most out of a well, a company needed to call in a "competent HydraFrac expert," cajoled the ad. A 1956 thesis on the subject offered different advice: "To prevent unwarranted expense and failure, only wells and equipment in good condition should be selected, as high pressures will be encountered in the process. Remembering that the fracture will follow the line of least resistance, a good primary cementing job is essential to confine the fracturing fluid to the zone to be treated." 41

In 1957, the year Jessica Ernst was born, Marion King Hubbert, a Shell geophysicist later known as the father of peak oil, hailed hydraulic fracturing as "one of the major developments in petroleum engineering of the last decade." But the scientist, an expert on the movement of fluids underground, offered a prophetic warning about the technology's loose cannon— fracking out of zone: "It is obvious that vertical fractures will facilitate the vertical migration of fluids where the fractures intersect permeability barriers. They may in this manner interconnect several separate reservoirs in lenticular sandstones imbedded in shales, and may in fact tap some such reservoirs not otherwise in communication with the fracture well. There is a danger, however, where a reservoir is overlain by a thin permeability barrier and a water-bearing sand or sandstone that a vertical fracture may also permit the escape of oil and gas into the barren sands or sandstones above."

But hydraulic fracturing, the injection of fluids to crack rock, was just one of the new players in a technological revolution that included computers and satellites. Jacques Ellul, a French radical Christian philosopher, outlined the character of this transformation in his 1954 *The Technological Society*, published in English a decade later. His celebrated book explained how techniques ranging from televisions to propaganda (ways to engineer people's minds) had turned people into slaves of efficient machines. Ellul argued that technology, including the construction of complex cities, had become the dominant and determining factor in society. Technology had so transformed communications, elections, and travel that people could no longer live spontaneously any more than an astronaut could walk freely in space without a life-support system. Humans depended so heavily on this artificial world of technique, said Ellul, that they had lost touch with the natural world. Moreover, these technologies, whether pesticides or fracturing, tended to destroy or subordinate the natural world "and did not allow this world to restore itself."

He also argued that technology eliminated human choice, "because the machine world favors technique that gives the maximum efficiency." As a consequence, opposing technology was no longer considered an option: "If a machine can yield a given result, it must be used to capacity and it is considered criminal and antisocial not to do so."

Ellul warned that technologies such as propaganda would greatly modify society. Moreover, they would suppress any notion of subject, as well as meaning, because everything in a technological world "is a means and only a means, while the ends have practically disappeared." In this regard, he argued, all technology was basically amoral. Technology makes no distinction between good uses and bad ones. It tends "on the contrary to create a completely independent technical morality."

Ellul believed that the state played a pivotal role in the spread of technology. Technologies and their machines increasingly created social and environmental problems, but only the state had the money to fully research them, and its scientific solutions invariably called for more technology—which ultimately centralized more power. For Ellul, statesmen had become "impotent satellites to the machine," imposing the chief goal of service and utility on its scientists and researchers.

According to Ellul, the proliferation of technologies also destroyed any notion of responsibility. Consider the example of a dam, he suggests. One day the dam's walls fracture and burst. A community is flooded. Who is responsible for that? "Geologists worked on it," Ellul wrote. "They examined the terrain. Engineers drew up the construction plans. Workmen constructed it. And the politicians decided that the dam had to be in that spot. Who is responsible? No one. There is never anyone responsible. Anywhere. In the whole of our technological society the work is so fragmented and broken up into small pieces that no one is responsible. But no one is free either. Everyone has his own specific task. And that's all he has to do."

The radical thinker pointed to the history of atomic energy as another illustration. Having created such a powerful and destructive instrument as the atomic bomb, why had society deployed it so quickly? Ellul's answer was disquieting. "Everything which is technique is necessarily used as soon as it is available, without distinction of good or evil. This is the principal law of our age. We may quote here Jacques Soustelles' well-known remark of May 1960 in reference to the atomic bomb. It expresses the deep feeling of us all: 'Since it is possible, it was necessary.' "

During the 1960s, while Jessica Ernst was attending elementary school in Montreal, the North American oil and gas industry embraced "nuclear fracturing" as a technology that was both possible and necessary. The goal was to crack open difficult rock formations with bombs powerful enough to level two Hiroshimas. "Whereas a hydraulic fracturing treatment usually produces a single fracture through the wellbore, a nuclear explosion can be expected to form multiple fractures," gushed one industry treatise. As engineers saw it, the nuclear stimulation program promised to release endless fountains of fossil fuels.

Soviet technicians were the first to experiment with nuclear fracking, because they had few social constraints to worry about. Between 1965 and 1989, the Nuclear Explosions for the National Economy program fracked approximately twelve oil and gas limestone formations with 2 to 7 kiloton bombs and the glorious aim of "forcing the depths to give up their riches." At an oil field south of the city of Perm, radioactive gases from the Grifon site went on to infiltrate sixty-five conventional wells and contaminated groundwater. After another nuclear frack job in the country, radioactive gases migrated to the surface and contaminated a heavily populated area over a mile in diameter. A Russian academic later concluded that nuclear fracturing never achieved commercial popularity in the Soviet Union due to "contamination problems encountered at the Grifon field." There was also public resistance "to accepting a product containing any added radioactivity."

American engineers also cracked tight shale rocks with nuclear bombs, but on a more modest scale. In 1967, Project Gasbuggy fractured a 4,000-foot-deep formation in New Mexico. Two years later, Project Rulison, in Garfield County, Colorado, set off a device three times more powerful than the bomb that had devastated Hiroshima. The Gasbuggy frack job created an underground hole 89 feet wide and 335 feet high that filled with radioactive gas. No public utility found a market to sell the stuff. Project Rulison, with a 43-kiloton bomb, exploded at a depth of 8,000 feet in tight shale, damaged homes miles away, and produced gas that was also radioactive.

(Although the U.S. government promised at the time that it would keep oil and gas drillers three miles away from the Project Rulison site, landowners and politicians had to fight bitterly to prevent industry from fracking into the radioactive rubble.)

    45

Not surprisingly, Gasbuggy and Rulison convinced U.S. researchers that nuclear stimulation presented a technical botheration: "Radioactive contamination of hydrocarbons is a problem with nuclear well stimulation." Undaunted, engineers predicted that "successful utilization of nuclear explosions" would someday "materially increase the world hydrocarbon reserves." They weren't far wrong. The exponential expansion of energy needed for hydraulic fracturing was in the end delivered by huge diesel trucks equipped with powerful pump engines. By 2006, industry would be using the equivalent of eleven nuclear reactors to deliver 11 million horsepower to rubblize rock in multistage hydraulic fracturing sites across North America. The technology marched on and broke new frontiers.

U.S. government scientists also came up with the idea of applying fracking technology to geothermal energy. Experts knew how to mine steam from deep rocks, but such deposits were rare. Most hot rocks deep in the earth held no water or steam. To investigate the possibilities, Japanese and U.S. researchers decided to drill two wells at the base of a couple of volcanoes to create a hot dry rock heat mining system. The plan was that one well would inject thousands of gallons of water into the volcanic rock and the other would pump the heated water back out. To create a connection between the two wells, the researchers proposed to frack the hell out of the granite. But nothing worked as planned. Researchers had hoped to create a single vertical fracture in the hot rocks, but the fracking treatment opened preexisting joints in basement rock instead, which reached out in every direction.

In the 1960s the U.S. military also learned something new about fluid injection: it could cause earthquakes. Outside of Denver, the Rocky Mountain Arsenal had decided that that best way to deal with a leaky waste pile of chemical weapons was to drill a 12,000-foot injection well and squeeze more than 17 million liters of hazardous waste into the earth's crust every month. But that activity provoked an epidemic of tremors. After army researchers matched injection rates with the earthquakes, they shut down the well. A year later three major earthquakes, each with a magnitude of 5, shook the region. The temblors split foundations, cracked walls, and broke windows in the Denver suburb of Northglenn. "Before this episode, the seismic hazard associated with deep well injection had not been fully appreciated," two U.S. federal geologists later explained.

In 1970, two Pan American engineers, G. C. Howard and C. R. Fast, summed up the state of the fracking business in what became a classic monograph. Their paper noted that technology had matured into bigger and bigger fracks. In its early days, hydraulic fracturing used 75 horsepower to propel about 420 to 840 gallons of fluid down a hole, making fractures about 10 feet long. Twenty years later, pumping trucks were delivering 850 horsepower to inject 25,000 gallons, breaking open rock with fractures ten times the original length. By the 1960s, injection rates had increased from 3.5 barrels per minute to 25 barrels, and by the 1970s, to 100 barrels.

Howard and Fast saluted the technology as an industry savior: "It has altered pipeline construction, changed production

46

practice, killed the nitro-glycerin oil well shooting business, upset the drilling business and revolutionized the service companies." By the 1970s, thirty different fracking companies were plying their fancy tools—services with names like Sand- frac, Riverfrac, HydraFrac, and StrataFrac—to the oil and gas industry the same way pharmaceutical firms pestered doctors with drug paraphernalia. The industry spent more on fracking services than it did on properly cementing wells.

Fracking had also changed the industry from a drilling operation into an increasingly complex engineering endeavor. Technicians reanimated exhausted wells through the application of more and more energy, equipment, water, and capital. A successful fracture job on an ailing oil well could ultimately retrieve an additional 17,750 barrels, reported Howard and Fast. Some stimulated wells increased their oil production twenty-fold. All in all, the researchers calculated, since the 1950s half a million frack jobs had probably added 7 billion more barrels of oil into the marketplace. (The United States consumes 7 billion barrels a year.)

Although the Pan American engineers described fracturing as a simple process, they acknowledged that "many important variable factors," such as fluid and pressure, ultimately influenced its cost and outcome. There were mysteries, too. Some frack jobs delivered more oil or gas than the industry-made fractures should have released. The engineers suspected these higher returns were due to fracturing "into zones of higher permeability" or to forcing cracks into other formations.

The report admitted that there was limited data on fracture depth and orientation. Moreover, fracking materials and techniques had improved much faster than "the understanding of the mechanics of in-situ fracture of the actual rock formations."

In other words, hydraulic fracturing remained a glorious science experiment that just happened, in many cases, to improve well production.

Howard and Fast's monograph mostly avoided the subject of groundwater. But the engineers recommended great care be taken to "prevent inadvertent fracturing into zones containing undesirable fluids," such as an aquifer. They warned that "injected treating fluids will follow the path of least resistance. No oil production increase can be expected if a fracture is created in cement, shale or coal, instead of the producing zone." Vertical fractures, they wrote, "can be created unintentionally and can extend many feet into water-bearing formations." The paper added that fracking beyond the pay zone, a common occurrence, was "not compatible with fracturing."

Patents filed with the U.S. government in the 1960s, '70s, and '80s portrayed a wealth of subterranean fracking worries. Almost all tellingly referred to fracking as an "art." Everyone in the industry had trouble controlling the length and direction of their fractures. (The patents sometimes compared rock fractures to "the fracture that extends through a wooden log as a wedge is driven into it.") As a 1961 Dow Chemical patent revealed, "No effective means or method is known to restrict or limit the cracks and fissures, produced during the fracturing operation, to the oil-producing horizons and thereby prevent their extension to adjacent water zones." In 1974, Mobil Oil's patent 3851709 A called uncontrolled fractures "certain" trouble: "When, for example, a gas-bearing formation overlies a liquid hydrocarbon-bearing formation or a water-bearing formation underlies a liquid hydrocarbon-bearing formation, it is undesirable for the fracture to extend vertically into either the gas-bearing or water-bearing formation."

A 1983 Mobil Oil patent filing (EP 0137578 A2) repeated the lament, citing "the undesirable flow of fluids from adjacent formations." The patent went on to describe the hitch as a severe limitation to the technology: "In such cases the resulting frac-   49
ture grows into the non-productive bounding beds, and some of the valuable fracturing materials are wasted. In cases where permeable beds containing unwanted fluids, such as water, are also penetrated by the fracture a large amount of unwanted fluid is introduced through the fracture into the producing well. In cases where the amount of such unwanted fluid is prohibitive, the well has to be abandoned."

Industry textbooks also highlighted wayward fracks as a concern. One text divided the underground world into positive and negative "fracture containment conditions." Positive conditioning included frack jobs in deeper formations, where out-of-zone fracks couldn't do as much damage. In shallow and difficult reservoirs where "negative conditions" prevailed, the textbook said, "excessive fracture height growth can only be prevented by renouncing from large-scale treatments which would waste fluids and proppants." The book noted too that excessive pressure often caused fractures to penetrate "overlying water- or other type of hydrocarbon-bearing horizons or... the ground water storey in case of very shallow reservoir depth."

These predictable problems didn't stop the industry. During the 1970s, frackers kept experimenting with different fluids, materials, and methods to accommodate deeper vertical wells. The number of chemicals used in fracking fluids to reduce friction, kill bacteria, or limit scaling in wellbores multiplied by the hundreds. Every year, industry filed an average of fifty fracking patents. One major innovation included the "use of metal-based crosslinking agents to enhance the viscosity of gelled water-based fracturing fluids for higher-temperature wells... The chemistry used to develop these fluids was borrowed from the plastic explosives industry." Stanolind tried rocket fuel as a fluid to crack open rocks, but with dismal results. The explosion left a five-foot-deep crater in the ground and eight workers dead.

The range of particles (proppants) flushed underground to keep the fractures open also expanded. The industry had started with river sand but soon graduated to ceramic pellets, high-strength glass beads, aluminum balls, walnut hulls, rounded nut shells, apricot pits, resin-coated sands, sintered bauxite, and fused zirconium. The proppant industry grew so rapidly in the 1970s, one textbook explained, that two-thirds of the cost of every frack went toward materials to keep the fissures open.

Inspired by a grandiose vision of fracking's success, the U.S. Department of the Interior and its Energy Research Program decided to up the ante. In 1976, government researchers declared that nuclear fracking wasn't effective enough. What was needed was the ability to target "gas in formations having permeabilities so low that gas cannot be produced economically." Only three technologies could do that: massive hydraulic fracturing (MHF); a combination of hydraulic and chemical explosive fracking ("subsequent detonation of the explosive generates high-pressure gases that expand and extend the original fracture"); and the fracking of deviated or slanted wells. Industry couldn't afford massive hydraulic fracturing jobs, the scientists concluded, but government could: "It is therefore justifiable to test the massive hydraulic fracturing concept, now thought to be a possible competitive (or companion) technique to the nuclear explosive method already being tested." One of the first MHFS, performed by government scientists in East Texas, deployed 340,000 gallons of gelled water and between half a million and a million pounds of sand.

By now the oil and gas industry had set off many earthquakes by lubricating faults with injections of water or briny waste into the ground. The first recorded earthquakes caused by big frack jobs shook up Oklahoma. In 1978 the "stimulation" of one 3,000-foot-deep well triggered 70 earthquakes in just 6.2 hours. The following year a massive frack experiment injected fluids at a rate of 230 litres per minute over four days into depths of 3 kilometers. That forceful blast of energy set off a swarm of nearly 100 earthquakes. Geologists later admitted, in a 1990 paper, that they had had no idea how big an earthquake fluid injection could trigger: "The process once started, may not be controlled completely or easily." Nor did they understand how earthquakes would affect the whole purpose of fluid injection—the containment of hazardous waste in the ground: "The occurrence of small earthquakes near the bottom of a deep injection well may indicate faulting or fracturing processes that conceivably could lead to a breach in the overlying confining zone and therefore conceivably could permit hazardous materials to migrate upward toward potential drinking water supplies." But there was now no doubt that fracking made earthquakes.

During the 1980s, when Jessica Ernst moved to Alberta and started to work in the oil patch, researchers basically declared war on permeability, a fancy measurement that tells you how easily liquids and gases move through a rock. Until the 1970s, most gas drilled in North America collected in formations almost as permeable as beach sand. Draining such pools required a well-positioned well that recovered 60 to 80 percent of the fossil fuels there. But as these cheap and easy sources dried up, the industry was faced with mining extremely deep or shallow formations. These complicated deposits weren't very permeable. Some shale gas formations were as tight as a building made of stone or bricks. Others were as impermeable as concrete or granite.

The decision to go after these extreme resources changed the nature of oil and gas extraction. Industry had to plant more wells to crack open more rock—rock that sprawled over a vast underground universe. The application of additional brute force, proppants, and fluids secured fewer molecules of hydrocarbons at greater cost than extraction from discrete conventional pools. Extraction from permeable formations required four vehicles, 20,000 gallons of fluid, and 100,000 pounds of proppants, often apricot pits or sand. Extracting oil and gas from impermeable rock took an army up to fifty times greater in size: forty vehicles, a million gallons of fluid and 3 million pounds of sand. Researchers dreamed that fracking with vast volumes of highly pressurized water would make the fishing expedition more rewarding. They wanted to frack not only shale and tight sands but coal formations too. Without much fanfare or public debate, the oil and gas industry was about to become a bottom-trawler operation.

It would take the industry more than a decade to figure out slick-water fracturing: "a fracture treatment that utilizes a large volume of water to create an adequate fracture geometry and conductivity to obtain commercial production from low permeability, large net pay reservoirs." In the meantime, it bombarded coal seams with a variety of complex experiments. The proposed fracking of coal beds sparked immediate

litigation in Pennsylvania. The U.S. Steel Corporation owned
a coal field there that it planned to mine one day. The corporation assumed it also owned the methane gas in the seam,
but a local oil and gas driller thought differently. In 1978, he 53
proposed to stimulate the eight-hundred-foot coal seam with
"hydro fracturing." To extract the methane, the driller planned
to create fractures "in the coal vein" that would serve as channels through which gas would flow to the wellbore. Lawyers
for U.S. Steel argued that hydro-fracturing constituted trespass on their property and would do irreparable harm to the
coal. Judge Glenn R. Toothman ruled on the high-profile case
in 1980. He agreed that the gas pulled from vertical wellbores
belonged to the driller, but he forbade the driller from fracking
the seam.

In his decision, Judge Toothman described the otherworldly makeup of coalbed gas. It is, he wrote, "similar to other
natural gases found below the earth's surface in composition
and content, but... in the manner of its origin has, different
from the other gases, a close affinity for and association with
coal seams. In its original state it permeates and penetrates the
coal bed, is its alter ego, its constant companion, its geological
handmaiden, and is sometimes viewed as its contumacious
free-spirited bride, but more generally regarded as its ill-
chosen bridesmaid. It is found with the coal when they come
to mine it, stays with the coal as it leaves, and remains in the
space after the mining has been done. Its past has been filled
with peril and tragedy, its present is seen as having a modest
commercial attractiveness, and its future as a fuel potential has
become increasingly brighter."

More litigation would follow, and the complexity would
remain. In 1981, participants at a resource engineering

workshop for the U.S. Department of Energy's Methane Recovery from Coalbeds Project compiled a list covering what they didn't know about fracking coal seams. The list went on for several pages. Researchers didn't know what caused poor frack results in coal. They knew little about the behavior of proppants or fluids in coal seams. Other than by seeing gas flowing, they couldn't tell if and how a frack job had actually worked. The size of reserves and the rate of recovery were usually unknown. There were no computer models or design programs for fracking coal beds. Scientists didn't understand the mechanics of coal seams as methane machines. "What is the aim of a frack—do we always want to stay in the coal seam?" one participant asked. Struggling with that question, researchers answered that if "an aquifer exists above the roof rock, you would not want to disturb it."

After two decades of wild experimentation in the United States, in 2001 the pioneering industry quietly invaded Alberta. Until that time, Jessica Ernst knew nothing about the technology of fracking. Although she had worked on the impact of drilling and pipelines on wildlife and rivers, and had promoted consultation with affected parties, industry's plans to crack open coal, one of the world's most fractured rocks, caught her by surprise.

It also unsettled a great many Americans . . . and their water wells.

FOUR

Before Shale:
Coal

IN THE MID-1980S, Colorado rancher Laurel Lyon found herself in the middle of an industrial experiment: the fracking of coal seams. Shattering coal beds over a large area had never been done in the United States, and nobody knew how well such an experiment would work. But government scientists and industry geeks were keen on trying it. They suspected that the methane would flow freely—and that the financial rewards would be immense.

Lyon's small ranch sat on top of the Fruitland coal formation in the San Juan Basin in the southwest corner of Colorado. The basin, a dramatic landscape of arroyos and pinyon trees, had been thoroughly drilled for oil and gas since the 1920s. But the Fruitland coal fields, despite some tests in the 1950s, had been largely untouched. Soon, though, the San Juan would become one of the most heavily fracked landscapes and most prolific gas fields on the continent. Industry would later describe the basin as "the Persian Gulf of Gas."

Lyon knew little about industry's ambitions, let alone the technology of fracking. But she did have a thorough appreciation of the state of her water well, which sustained her pigs, chickens, horses, and goats. Prior to fracking, her water was wet, safe, and drinkable, even though "it wasn't as tasty as spring water." Three months after Amoco drilled a coalbed methane (CBM) well about a thousand feet outside her sixteen-acre ranch, all that changed. Lyon's water now contained a new ingredient: methane. She noticed the transformation while filling stock tanks—the water fizzed like soda pop. The taste soon got so bad that Lyon stopped brushing her teeth with it. To her consternation, she even discovered that she could light her faucet on fire. Nearly a half-dozen of her neighbors, it turned out, could perform the same trick.

Then came more trouble: Lyon's horses and goats started to abort in unprecedented numbers. Milk production in her goats declined by half, and the animals couldn't suckle their young. Alarmed by the deteriorating health of her livestock, Lyon performed her own science experiment. She used water from her 340-foot-deep well to irrigate one part of her garden, and used only hauled water to nurture another plot. The roots of the plants fed by the contaminated well "got fried," while the corn and squash irrigated with trucked-in water thrived. Lyon, the daughter of a Los Alamos snake venom researcher, photographed and documented the results. To this day, she suspects a friction reducer (sodium chloride) in the frack fluids acted as a protein inhibitor and sickened her livestock. She kept sacks of approximately forty aborted livestock fetuses in the freezer as evidence.

At first, industry and the Bureau of Land Management (BLM) gave Lyon the standard brush-off: they claimed that the methane came from natural sources. Lyon knew there were lots of natural gas seeps in the area, but her water hadn't gone bad until Amoco drilled and fracked the coal seams. Next, industry technocrats and government bureaucrats showed the landowner illustrations of San Juan geological formations that looked like pancakes stacked on top of each other, explaining that there was no way gas could travel through all these formations and penetrate shallow aquifers. But Lyon knew the geology at the base of the Rocky Mountains was highly folded and therefore complex. She laughed at the textbook geological maps and said they didn't portray reality. Industry then tried to blame the groundwater contamination on her livestock. Lyon said that was nonsense: none of the livestock pens stood anywhere near her water well.

Lyon's carefully collected evidence eventually became the basis of a class action lawsuit involving more than seventy landowners in Colorado and New Mexico. The lawsuit charged seven oil and gas companies, including Amoco, with operating CBM wells that caused air, water, and soil contamination. One of Lyon's lawyers, Ed McCord, even obtained Amoco diagrams that showed how drilling in coalbed seams could allow methane to escape to the surface or into the groundwater. In 1993 McCord told *High Country News*, "County officials and many local residents suspect that the pumping [of water] may have changed underground water pressure across large areas, freeing gas to migrate through natural cracks and fissures and residential water wells." He pointed to a 1989 internal BLM memo warning that CBM drilling activity "may be causing new leaks in the estimated 15,000 old and abandoned wells scattered across the basin."

Amoco's lawyers hogtied the lawsuit in a jurisdictional dispute about which federal court should hear the matter. Because the majority of the landowners lived within the boundaries of the Southern Ute Indian Reservation, industry lawyers argued, the plaintiffs had filed their negligence suit in the wrong court. As a consequence, explains Lyon, "We never got to the point of showing the evidence in court." The oil and gas companies knew "what evidence we were sitting on and didn't want a judge to see it," she says.

No judge ever did. Fearing that the jurisdictional issue might cost them the case altogether, the landowners settled out of court in 1996. As part of the settlement, the oil and gas industry insisted the landowners sign a three-year nondisclosure clause. Lyon was one of the last holdouts. She highlighted the most offensive parts of the confidentiality agreement in ink that she made out of bull manure, and mailed it back to the companies. To appease other plaintiffs, she eventually signed the "tiniest signature" she could muster. Immediately afterward, Amoco stopped delivering clean water to her farm.

Lyon says today that she learned several things from the experience: "The industry sucks. It doesn't care about neighbors. And all it cares about is money." She adds that eighteen years later, she's still paying for Amoco's industrial experiment: "I'm still hauling my water."

TWO DECADES BEFORE the fracking of shale rock would create a storm of rural protest and hundreds of documented cases of groundwater contamination throughout North America, government subsidies and research funded a coalbed methane revolution in the United States. The unconventional assault on coal, which Halliburton would later call "production enhancement," dramatically unsettled communities from Colorado to Alabama. The aggressive industry would eventually give birth to a methane cloud half the size of Connecticut, which now hovers over the Four Corners region of the country. In 2014, scientists called the cloud the largest "methane anomaly" that can be viewed from space; the hot spot represents nearly 10 percent of the United States' methane emissions.

CBM, the oil and gas industry's first blitzkrieg on unconventional resources, remains a startling treatise on the hazards of hydraulic fracturing. The original aim was to extend the shelf life of fossil fuels by exploiting more difficult and extreme resources. But neither industry nor regulators appreciated the complexity of coal or the consequences of fracking one of the world's most naturally fractured rocks. The smashing of coal seams in nearly a dozen basins left a troubling legacy of exploding homes, flammable water, uncontrollable methane seeps, and earthquake swarms, as well as major lawsuits in Alabama and Colorado. The frenzied operation also led to something called the Halliburton Loophole, a lobbying effort to exempt the U.S. fracking industry from federal oversight and the Safe Drinking Water Act. As such, the United States' CBM revolution offered an uncanny preview of the political controversy and geological unrest that the technology of fracking would unleash as it spread to other unconventional resources. It also serves as a damning prologue to Jessica Ernst's story.

Long before the oil and gas industry liberated methane into groundwater, the odorless gas had an explosive history: it blew up coal miners. Coal, the world's most abundant fossil fuel, has been mined and burned for centuries. It still provides nearly a fifth of the world's energy. But digging up 200-million-year-old plants compressed into a brittle and highly fractured material has always been a dangerous business. Early bacterial action on peat, combined with millions of years of pressure and heat, makes coal both a source and a reservoir of gas. The gas not only sits in the coal but strays to formations around it. Volume for volume, the highest rank coal (the well-aged and cleaner-burning anthracite coal) can hold more methane than a sandstone formation. But the combustible ancient rock remains a schizophrenic substance. It is a fuel, a rock, a solid colloid, a chemical reactant, and an organic sediment all in one. As a consequence, one 1990 conference paper on CBM observed, it is not surprising that "the reservoir behavior of coal [is] so difficult to predict and model."

In *The Road to Wigan Pier*, his 1937 book about depressed mining regions in northern England, George Orwell emphasized two other truths about coal. The first was that energy consumers generally remain oblivious to the dirty, complicated messes created by mining energy. "Practically everything we do, from eating an ice to crossing the Atlantic, and from baking a loaf to writing a novel, involves the use of coal, directly or indirectly," wrote Orwell. "For all the arts of peace coal is needed; if war breaks out it is needed all the more. In time of revolution the miner must go on working or the revolution must stop, for revolution as much as reaction needs coal. Whatever may be happening on the surface, the hacking and shovelling have got to continue without a pause, or at any rate without pausing for more than a few weeks at the most. In order that Hitler may march the goose-step, that the Pope may denounce Bolshevism, that the cricket crowds may assemble at Lords, that the poets may scratch one another's backs, coal has got to be forthcoming. But on the whole we are not aware of it; we all know that we 'must have coal,' but we seldom or never remember what coal-getting involves."

Orwell's second key revelation concerned coal's extraction: "Coal lies in thin seams between enormous layers of rock, so that essentially the process of getting it out is like scooping the central layer from a Neapolitan ice." This constant scooping by early coal miners acted on the coal much like a primitive horizontal drilling fracking crew, without the extreme horsepower.   61

The similarities between early mining techniques and contemporary fracking are profound. Nineteenth-century miners first sunk a vertical shaft into the coal seam, extending it several hundred feet below the ground. Then they branched out to claw the coal from long horizontal tunnels as they followed the coal seam. Next came the splitting and cracking of coal cleats. Miners didn't use highly pressurized fluids or sand to shatter the face of the coal but instead employed cudgels, pick axes, explosives, and, in the twentieth century, electricity-powered cutting machines.

All of that splitting and cracking released copious amounts of methane. Among the cleats, or mini-fractures, that decorate the face of every lump of coal dance molecules of methane. (The face of coal contains ladder-like fractures. The vertical strands are known as *face cleats*, and the horizontal cracks are called *butt cleats*.) The gas not only hovers around the coal but sticks to it, like sweat to skin pores. Coal's unique properties make it both a "gas generating machine" and a sponge. By shattering the wall of a coal seam, miners break the molecular chain and free the adsorbed methane. They also liberate the explosive gas in another way. Coal seams routinely act as pipelines for the movement of groundwater. In fact, much drinkable water travels through coal and saturates its seams. The water pressure, however, keeps the methane locked in.

Methane has many extraordinary properties. You can't smell, see, or taste the world's lightest hydrocarbon. It also behaves like some itinerant vagrant, escaping from the earth into the atmosphere or groundwater via a vast network of fractures and faults. The gas not only floats above air but strays from high- to low-pressure zones. It is the primary constituent of natural gas and, in the short term, a much more potent greenhouse gas than CO_2. Methane rarely travels alone. It often keeps company with other migrants, including radon, CO_2, ethane, propane, and hydrogen sulfide.

The earth creates methane biogenically from microbes and thermogenically from rocks. Ocean microbes first turned carbon monoxide into methane and vinegar billions of years ago. Now bacteria farts the gas from swamps, wetlands, landfills, rice paddies, and the stomachs of cows. Methane made by living things at shallow depths and temperatures below 80°C (175°F) has a distinct chemical, or isotopic, fingerprint. Methane forged by the heat (160°C to 200°C, or 320°F to 390°F) and pressure of deeper oil- and shale-bearing rocks looks much different. It contains heavier carbons.

Wherever methane accumulates, it displaces oxygen and asphyxiates life. An enclosed space that contains 5 to 15 percent methane and 12 percent oxygen is primed to explode. Methane that invades soils starves plant roots and robs them of oxygen. And over centuries, clouds of methane have suffocated miners or erupted into flame, leaving a record of subterranean carnage.

Coal miners regarded the water running through coal as a nuisance and an obstacle. James Watt modified the steam engine in the 1770s primarily to help early coal miners pump water out of flooded mines. In the process, the miners released so much methane that they turned coal mines into bombs. Soot-faced colliers feared methane's presence and had already long referred to the gassy hazard as "fire damp," an expression that originated from the German word for fog and vapors. Coal miners had many names for the dangerous gases that accompanied coal. Hydrogen sulfide, a brain killer, was "stink damp"; carbon dioxide, a life suffocator, was "choke damp"; and carbon monoxide, another smothering gas, "white damp." Miners called the vapors created by their breath and by the smoke of candles "globe damp." But for centuries, "fire damp" remained a constant foe.

As the Industrial Revolution exhausted shallow coal seams in the eighteenth century, the industry dug deeper, thereby encountering more fire damp and water. Steam-driven pumps removed the water, but the process released more methane. Lighter than air, the gas bubbled up from the shaft floor or hissed from coal fractures along the tunnel ceiling. Some mines were so gassy they supported a jet of flaming gas near their openings. Miners called these seeps "everlasting lamps." To ventilate the mines, children as young as five opened and closed trap doors all day long to break up concentrations of fire damp.

Methane mine explosions came in legendary types and sizes. Mild bursts could knock a miner flat or singe his hair. Massive conflagrations mangled and buried both horses and humans in a bloody tangle of timbers and coal. Fatal events, sometimes described as "rivers of fire," propelled bodies from the mouth of a mine as if shot from a cannon. Such conflagrations often sounded like a monstrous clap of thunder. In 1709, one English blast chucked two men and a woman from the mouth of a mine 57 fathoms (342 feet) into the air. They found one of the men headless, a frightful distance away, while the woman had "her bowels hanging around her heels."

Methane seepage, or "fulminating damp," so terrorized early British miners that the industry created the unique job of "fireman." These "men of purpose" dressed like wizards in special water-soaked cloaks made of leather and wool. Every day, they sought out and burned the accumulating pockets of fire damp that fizzed from crevice to crevice to make conditions safer for their brethren. Hooded faces close to the ground, firemen would raise lighted candles attached to sticks to the pockets of methane floating along the ceiling. Their dangerous handiwork awed a sixteenth-century visitor to an English coal mine: "The miner then draws near to the fire, and frightens it with his staff. The fire then flies away and contracts itself by little and little; having then expended itself, it collects itself together in a surprising manner, and becoming very small, remains quite still in a corner." Most firemen didn't live to become grandfathers.

But despite the best efforts of firemen, "the invincible enemy" of methane continued to kill more and more coal workers. The explosions happened so frequently in Newcastle, England, during the seventeenth century that authorities recommended that "we drop the further mentioning of it." Between 1870 and 1879, the U.S. government recorded 639 fire damp explosions resulting in 1,127 casualties.

In recent years, the coal-mining industry has sanitized the language related to methane disasters. For example, industry now refers to catastrophic methane flare-ups as "outbursts"—a violent simultaneous release of gas. A 2014 Elsevier textbook, *Coal and Coalbed Gas: Fueling the Future*, explains that coal mines have recorded 30,000 "outbursts" in the last 150 years. In plain language, these "outbursts" killed tens of thousands of miners and spewed volcanic amounts of methane into the atmosphere. Methane "outbursts" in Chinese mines still kill an average of one miner a week. Due to the complexity of underground stresses, the mechanics of coal fractures and the presence of gas-bearing faults, scientists cannot predict when a methane "outburst hazard" might occur.

With four catastrophic methane explosions in 1907 that buried or burned more than six hundred U.S. miners, the coal-mining industry cemented its global reputation as a "callous" and greedy institution. Newspapers described companies mining coal as "the most backward of the civilized nations." The carnage forced the U.S. government to create the U.S. Bureau of Mines, which was charged with developing safer ways to mine coal and ventilate dangerous gases. That's how the coalbed methane industry slowly began.

The Bureau of Mines first experimented with drilling holes in sandstone formations above gassy coal seams, to release fire damp into the atmosphere. But they couldn't keep pace with the machines that by then continuously chopped up coal beds, going much faster than men with picks. The machines fractured the coal so quickly that they accelerated the rate at which methane poured from coal cleats. As a consequence, even more workers died in sweeping explosions. The bureau later calculated that gassy coal that was "highly stressed and penetrated by mining-induced factures" was "outburst prone."

The bureau and its researchers kept experimenting. They tried drilling vertical holes 400 meters (437 yards) in front of advancing mines to vent the methane. They also drilled horizontal wells from the bottom of the shaft into virgin coal seams. These wells sometimes reduced the amount of methane released at the coal face by up to 40 percent. Nevertheless, the bureau thought it could do better using hydraulic fracturing.

In the 1970s, bureau researchers planted five vertical holes in advance of a new coal mine in Jefferson County, Alabama, then fracked the wells with gelled fluid and sand. After they had pumped out 40,000 barrels of water, the amount of gas produced by these experimental wells increased fourteen-fold. A 1974 U.S. government study gushed that the fracking of coal seams could "decrease the hazard of methane ignitions and explosions during mining" as well as provide "supplemental sources of fuel." To reassure the mining industry that hydraulic fracturing wouldn't undermine mine roofs or walls, the Bureau performed more tests at mine sites in Illinois and Pennsylvania.

These so-called mine-back demonstration projects offered a couple of surprises. At a Methane Recovery from Coalbeds Symposium held in 1979, researchers admitted they couldn't always predict where the fracks would go. Nor could they explain why half of the time nitrogen frack treatments extended through and beyond "strata immediately overlying the coal seam." They were at a loss to explain why some treatments created fractures more than ten times wider and shorter than predicted, but researchers suspected that the incongruous nature of faulted and fractured coal seams might explain "the cause of the considerable deviations between the predicted and observed fracture behaviors." Later federal government studies found that fracking fluids traveled six times farther than sand—anywhere from 200 to 600 feet. Fluids seemed to follow a stair-step pathway through the coal seam. Another repeated finding puzzled researchers: the quantity of methane released after fracking often exceeded calculations about what the coal contained. That meant some of the methane was traveling from neighboring formations through natural fractures in ways no geologist comprehended or could predict.

Although hydraulic fracturing and improved venting allowed the industry to collect and use methane from coal mines, methane still seeped from 8,000 abandoned mines in the United States at a rate of 13 billion cubic feet a year. The gas poured from old entrances, boreholes, venting pipes, and ventilation systems, as well as from "fractures in the overburden rocks." Active mines released another 174 billion cubic feet of methane from chopped-up mountain tops, ventilation shafts, and coal-handling facilities. Industry captured about one-third of that liberated gas. It couldn't seal all the leaks started by man-made fractures into coal seams.

From North Derbyshire, England, to Pennsylvania, communities that sat atop old coal mines regularly experienced household methane invasions as well as uncontrolled explosions. In an exhaustive textbook on coalbed methane, U.S. geologist Romeo Flores explained that mined coal seams were too fractured to fully contain the explosive gas: "Fracture systems caused by displacement and deformational stress and strain of the overburden rocks before, during, and after mining cannot be totally accounted by sealing man-made structures."

In the late 1970s, U.S. research abruptly changed its focus from mine safety to methane mining. Severe shortages of conventional natural gas had closed schools and public buildings and sparked acrimonious congressional hearings. Faced with dwindling domestic oil and gas supplies, the U.S. government decided to attack "poorly understood and high cost" resources, including shale gas, tight sands, and coalbed methane. Yet a landmark 1978 study warned that cracking open these unconventionals would be "controversial, costly and risky." In fact, at the time, the oil and gas industry considered the idea of sucking methane out of coal beds so preposterous that the resource was referred to as "moonbeam gas."

Most of the easy gas that drifted into pools above hydro-carbon-rich rocks had already been siphoned off, so industry set out to tackle the source rocks at the bottom of the energy pyramid, "in geologically complex, extremely low permeability (tight) reservoirs." Gaining access to these extreme rocks would require more money, water, and energy, but armed with over $200 million in federal research funds, the Department of Energy and its Unconventional Gas Recovery program started to probe and measure. Researchers attended seminars and established illuminating journals such as the *Quarterly Review of Methane from Coal Seams Technology*. Federal scientists hoped that hydraulic fracturing might be the Holy Grail for shale rock and coal.

Since the gas production rates associated with unconventional wells had so far been "too low to be economically attractive," the federal project was directed toward "developing fracture, stimulation, and drilling techniques to connect more gas-containing fractures to the wellbore." The resulting research, combined with the situation of dwindling domestic supplies, spawned an unlikely courtship between coal geologists and natural gas drillers. Their diffident marriage eventually produced the experimental and highly complex coalbed methane industry.

The first coalbed methane wells were installed in deep coal seams about fifty feet thick in Colorado's San Juan Basin and in thin, shallow ones in Alabama's Black Warrior Basin. Federal geologist Craig T. Rightmire predicted early on that drilling coal for gas wouldn't be easy. He noted that coals, which he described as thin, compressible, and highly chemically reactive, behaved differently than oil and gas reservoirs. He also believed, correctly, that each coal basin was unique and would

require an entirely different fracking technique. But the industry charged forward and fracked the hell out of coal. In 1983,
annual CBM production in the United States grew from nothing to nearly 6 Bcf (billion cubic feet) with about 165 pioneer
wells. By 1994, the industry would be extracting 85.1 Bcf from
more than 6,000 wells. Over the next five years, the industry
planted 14,000 CBM wells, producing roughly 9 percent of the
country's natural gas.

69

Halliburton, which had been fracking energy wells since
World War II, hailed the experimental fracking of coal seams
as a godsend. According to company literature, industry had
transformed a deadly mining byproduct into a friendly fuel;
methane in coal had gone "from hazard to environmental
challenge to energy resource." CBM, Halliburton said, also
promised "a clean burning fuel" that would increase gas
reserves, improve mine safety and reduce the amount of methane vented to the atmosphere. Schlumberger, which had begun
in the 1930s as a well-logging company, published similarly
ecstatic CBM propaganda.

As companies began puncturing and rubble-izing coal
basins, the U.S. Department of the Interior and its Office of
Surface Mines quietly released a prescient report on methane
seepage. The 2001 report offered insights on how to track and
investigate fugitive releases of methane in coal-mining regions.
It noted that human engineering "can also cause or accelerate
the release or migration of methane. Fracturing of the overlying rock strata above mines and the drilling of boreholes can
alter the pressure gradients and create open pathways for the
gas to migrate to the surface." The report's authors explained
that "methane travels through the rock pores and fractures.
The porosity and permeability of the transmitting medium

significantly affect the migration. Methane, being lighter than air, tends to migrate vertically. However, pressure gradients cause methane to migrate to areas of lower pressure. These migrations may be in directions other than vertical." Last but not least, the report said, natural faults and fractures "may also produce a conduit for gas to travel far from its source," and methane-contaminated groundwater, when pumped into homes, could create a flammable hazard. "Occasionally, the levels are high enough that a kitchen faucet can be ignited with a match."

Coalbed methane promoters ignored these hazards, talking up the benefits of CBM instead. From an industry perspective, there were many. Well-mapped coal seams lay under 13 percent of the United States. The seams didn't require deep drilling, because they sat much shallower than most oil and gas deposits. The boosters also declared that a cubic foot of coal stored about seven times more methane than did an equal volume of sandstone rock in a natural gas reservoir.

CBM did pose two significant challenges to the industry, though: land disturbance and water disposal. Unlike conventional wells, which drained one large pool every section, CBM required eight to sixteen wells on the same amount of land. To drain small amounts of methane from coal seams stretching over a large area, industry had to carpet bomb the countryside with wells. The high density guaranteed that fracture treatments would purposely connect to other fracked well sites. It also guaranteed that the industry's presence on the landscape would expand with military precision and scale.

Water posed its own set of problems. Most coal seams serve as water pipelines or aquifers, and pumping off millions of barrels of water from coal seams would cost money. Disposing of these subterranean rivers in unlined ponds or injecting the water back underground into deeper formations would cost more money. That could make CBM largely uneconomical.

U.S. federal subsidies solved that conundrum, however, and sparked a frenzy of experimental drilling. After the oil price crunch in the 1970s, the U.S. government introduced the Crude Oil Windfall Profit Tax Act. Under Section 29, the act provided tax credits to companies trying to unlock unconventional hydrocarbons such as shale gas and coalbed methane. Between 1979 and 1989, about $270 million of credits went to companies drilling in Alabama's Black Warrior Basin, while close to a $1 billion helped companies puncture the San Juan Basin with thousands of wells. Another $900 million fueled the U.S. industry until 1993. In this frenzy to claim the credit, industry geologists gave little thought to how dewatered coal seams and man-made fractures might connect with natural fractures to release methane all over the place. It was landowners in Colorado, West Virginia, and Alabama who first found "damp gas," in their drinking water.

The picturesque San Juan Basin straddles the Colorado and New Mexico borders. Once home to the Ute people, Spanish friars, and sheepherders, in the 1920s and 1930s the basin became a province of the oil and gas industry. Between the 1980s and the mid-1990s, the industry targeted a part of the basin known as the Fairway and drilled some 2,000 vertical wells into the Fruitland coal formation. The wells varied in depth from 500 to 4,000 feet. Many targeted coal seams nearly 40 feet thick. After sucking out almost 30 million barrels of water a year in some cases, industry championed the San Juan as the highest gas-producing coal basin in the world. Some wells poured out between 1,000 and 15,000 cubic feet of methane a day.

To shatter the San Juan coal seams, frackers pumped down 55,000 to 300,000 gallons of fluids and 100,000 to 220,000 pounds of sand proppant per well. The frack fluids included hydrochloric acid, slick water using unknown solvents, nitrogen, CO_2 foam, and gels containing formaldehyde. As the U.S. Environmental Protection Agency reluctantly reported in 2004—nearly a decade later—almost every well in the San Juan Basin was "fracture-stimulated, using either conventional hydraulic fracturing in perforated casing or cavitation cycling in open holes." (Cavitation, a unique form of fracking, operates by blasting foam, air, or water into a coal seam until the mixture explodes, much like shaking a pop bottle.) In the San Juan Basin the cavitation of just one well vomited 100 tons of coal and water to the surface and created a hole full of methane-freeing fractures. As researchers admitted years later, containing the fracks in the targeted coal zone was difficult. In some places no shale rock provided a roof for the coal. As a consequence, hydraulic fractures in the San Juan Basin often fingered out into overlying beds of sandstone and groundwater.

Nearly a decade after the CBM boom, the BLM revealed that "several environmental situations of concern" had sprung up in "the progressive center of coalbed methane development from the Fruitland Formation." Gas and hydrogen sulfide seeps had appeared where none previously existed. Healthy springs had dried up. Older, natural seeps gushed poisonous gases in greater volumes than ever before. Underground coal fires broke out as industry dewatered some seams and introduced oxygen. Old gas wells served as pathways for methane migration into groundwater. Soils became so saturated with fire damp in the Animas River Valley, south of Durango, Colorado, that plant and grass roots suffocated. The zones of methane-killed vegetation extended for miles.

The startling BLM report, issued in 1999, also admitted that the federal agency responsible for the management of public lands had witnessed a dubious science experiment in the west. Basin geology was so complex and full of faults that it was impossible to predict with accuracy where the methane might pop up next. Moreover, said the report, the hydrology of the area was poorly understood: "Years may pass before a full understanding is achieved. The capability does not currently exist to predict the next area a problem might arise or to mitigate an existing seep."

As had rancher Laurel Lyon, Carl Weston witnessed methane liberation firsthand. After frackers invaded the Animas Valley, gas infiltrated the drinking water on Weston's property outside of Durango. Methane spit out of his tap in such forceful quantities that he abandoned his home in the early 1990s and moved into a mobile trailer on his property. The problem continued, and the contaminated water turned Weston's laundry the color of ash. "When you live around here, you've got to keep a special set of white underwear to wear to the doctor's," he told the *Denver Post* in 2001. "For a while I was a poster child for flaming water."

Near Weston's property, the BLM found an old gas well that had been plugged with a telephone pole. The well ran right through the Fruitland coals and became a conduit for methane. The bureau later spent hundreds of thousands of dollars plugging leaky abandoned wells in the area, but, says Weston, they were "only partly successful." To this day, he hauls fresh water to his trailer.

Methane seeps also appeared in rivers and creeks where landowners had not seen such things before. By the 1990s, La Plata County residents of a small subdivision by the Pine River reported that their kitchen sinks had become fountains of sparkling water supersaturated with methane. Some residents could even light the river running by their houses on fire. Large ponderosa trees started to die off due to methane suffocation, and state investigators found one house filled with so much methane that it stood on the verge of exploding. The residents traced their problems to ten coalbed methane wells that Amoco had drilled nearby in the late 1980s.

When La Plata County asked the Colorado Oil and Gas Conservation Commission to shut down the Amoco CBM wells, the commission initially refused. So did Amoco. The commission—a seven-person governor-appointed panel—also balked at doing a basin-wide environmental impact assessment of CBM. Not surprisingly, four of the seven commission members owed their living to the oil patch: a geologist, a geophysicist, a petroleum engineer, and an oil and gas attorney.

Later studies did find industry responsible for the seeps, and Amoco settled a lawsuit out of court. The company bought nearly half a dozen homes in La Plata County and bulldozed them. Amoco was also later forced to close one of its CBM wells. A real estate study commissioned by La Plata County found that having a CBM well in your backyard could diminish property values by as much as 22 percent. Despite all of this, Amoco never admitted to any wrongdoing.

CBM extraction in Colorado likely triggered a swarm of earthquakes, too. Three hundred and thirty-four tremors occurred near ten injection wells near Trinidad, Colorado, in the fall of 2001. The wells had been created to dump wastewater sucked out of coal seams into formations over nine miles deep. Although the quakes happened at comparative depths and the epicenters were all near injection wells, investigators couldn't prove conclusively that industry had triggered the quakes with fluid injection. In 2002, the U.S. Geological Survey played it safe: "The fact that the data do not make a strong case for the earthquakes being induced does not imply that the data do make a strong case against the earthquakes being induced." But the earthquakes that later dogged the fracking of shale formations in Texas, Oklahoma, Ohio, Alberta, and British Columbia suggest that the Colorado quakes were but a preview of things to come.

75

So many landowners complained about polluted water wells after the CBM drilling boom in the late 1980s that U.S. Representative Ben Nighthorse Campbell started a committee to study the affair. The group got the U.S. Geological Survey to investigate the occurrence of methane in groundwater along the Animas River Valley in 1990. The survey took four years, cost a quarter of a million dollars, and was funded by the Colorado Oil and Gas Commission, industry, and La Plata County. The chief investigator, Daniel T. Chafin, a shy, bespectacled man, thoroughly reviewed previous studies on gas migration, and he discovered that the oil and gas industry had a troubling record. In the 1960s, for example, researchers had found explosive amounts of methane in groundwater in southwestern Louisiana, near leaking natural gas wells. A leaky gas well in Ohio had contaminated groundwater with methane as well as iron, manganese, calcium, and sulfide.

After sampling methane from 205 water wells, nearby gas wells, and natural gas seeps in the Animas River Valley, Chafin discovered that CBM activity had provided a subterranean methane release. Both dewatering and fracking had allowed thermogenic gas to migrate from coal seams up existing industry wellbores and into sandstone formations. From there, the buoyant gas had strayed into natural fractures just a few hundred feet from the surface and into groundwater and the Animas River. Chafin reported that 70 of the 205 wells contained explosive amounts of methane. Wells highly contaminated by methane also contained large amounts of H_2S (hydrogen sulfide), a highly toxic gas. The "manmade migration pathways" for methane included hundreds of leaking conventional wells and uncemented gas wells along coals in the Fruitland Formation. Old seismic test holes also played a role. Chafin concluded that the high content of gas in soil around numerous gas wells showed "that the cement around many surface casings does not effectively prevent upward migration of gas from the annuli of those wells."

Both industry and the BLM attacked the study. They argued that the gas seeps in the Animas Valley area predated drilling and that methane had moved through natural fractures to local groundwater. They characterized Chafin's conclusions about "manmade pathways" as bad science. "We are somewhat disturbed that several apparent contradictions are present and many conclusions are drawn based on what could arguably be characterized as inconclusive data," wrote BLM manager Sally Wisely. "We are also concerned that, to a certain degree, the tone of the document seems to lack objectivity."

Although Daniel Chafin's report is now cited in almost every major scientific paper on leaky industrial wellbores, the U.S. Geological Survey kept the paper off the Internet for years. Incredibly, the EPA did not include Chafin's findings in its 2004 report on hydraulic fracturing in coal seams. In fact, the Animas study pretty much fractured Chafin's career at the USGS. After taking early retirement in 2004 to end "a 14-year nightmare," he made a curious video called "My Struggle with Tyranny." During the hour-long recording, the hydrologist read from a statement. He claimed that senior administrators not only tried to delay the Animas study but also kept a secret file on him. He claimed that either administrators or industry illegally tapped his phone at the same time. Chafin also expressed concerns about the USGS receiving funds from the natural gas industry and the Gas Research Institute. Toward the end of his video, he warned that "rapacious corporate feudalism" posed a grave threat to the future of U.S. democracy. "We must turn back this tide of fascism."

BY 1999, THE BLM had replugged or recemented nearly three hundred leaky wells. Many had proved so difficult to seal that the BLM had to reenter several times. Repairing the broken wells lowered the levels of methane found in groundwater in some parts of the basin but made little or no difference in others. The agency calculated that a minimum of 1 to 14 billion cubic feet of gas per mile could be vented to the atmosphere over the next two hundred years due to drilling and fracking in the San Juan Basin, noting in an understated fashion that "this is a significant amount of resource potentially lost to the atmosphere."

In neighboring Wyoming, coalbed methane drilling imposed a similar footprint. Throughout the famed Powder River Basin, coals lay under large and generous aquifers. Ranchers were the first to notice the complications as the number of CBM wells jumped from 1,000 to nearly 8,000 between 1998 and 2001. Industry pumped so much water from the basin that it cheekily advised ranchers to get out of the cattle business and start catfish farms. They plunked down wells every 40 acres, scarring arid historic landscapes with seismic lines, roads, and pipelines. Researchers later admitted that drilling in the basin was dogged by "incomplete critical information and incorrect interpretations" about where the groundwater flowed.

78   Ed Swartz, a third-generation rancher, presented a vivid picture of the ugly development to the U.S. Congress in 2001. The cowboy spoke bluntly. "There has been nothing orderly about this development, with the possible exception of the collection of revenues," he told the politicians. "While I and fellow ranchers have faced bad economic times, drought and other mining booms, nothing has presented the kind of challenges and damaging impacts to our soil, water and lifestyle as the CBM development."

Swartz tallied the consequences: the destruction of roads by industry traffic, the proliferation of noisy compressor stations, the scarring of the land with pipelines and noxious weeds. Most disturbingly, he outlined how the industry discharged 44 million barrels of water a month, at 42 gallons per barrel, onto the land or into the ground. "That is enough water to supply 300,000 people per day, or ⅔ of Wyoming's population, or 2.5 million cows per day. This is water that we are pumping out and essentially throwing away."

Swartz then asked the question that landowners from shale gas fields of Pennsylvania, Ohio, Louisiana, and Texas would ask politicians a decade later: "Where are the protections for those of us bearing the brunt of the impacts for the development of this energy? The extraction of coalbed methane development is mostly experimental and the Powder River Basin has actually been referred to by industry representatives as a laboratory. Why should we, who call this place our home, be guinea pigs? We are watching our homes and ranches transformed into an industrial gas field. There

are about 14,000 CBM wells permitted in Wyoming's Powder River Basin and upwards of 75,000 to 100,000 wells
projected. The development of CBM is primarily being carried
out on the backs of landowners that have essentially no say   79
in how the development can proceed and are being required
to sacrifice our ranches, our water resources, our soil, our privacy, the wildlife—which also provides an income to many
landowners—and our livelihoods. And yes, it is clean energy
for urban areas—unfortunately, the production end in rural
areas is not clean. We should not let the glitter of ten to twenty
years of affluence blind us to the impacts and damages being
felt very directly by others that are not reaping the benefits."

As CBM turned much of the U.S. west into "sacrifice zones,"
the industry spurred a remarkable legal battle in Alabama.
The Black Warrior Basin, which lies in the west-central part of
the state, is the largest coal formation in the South. Until
1980, the Black Warrior boasted no gas wells. By 1996, more
than 5,000 vertical wells decorated the formation, which is no
thicker than a foot in places. Industry shattered the coal seams
with water, acids, diesel fuel, and liquid gels. Some CBM wells
removed as much as 30 gallons of water a minute from the
coal seams. Industry trucked the unwanted stuff to federally
licensed Class II injection wells that flushed the water into formations as deep as 10,000 feet.

During Alabama's CBM boom, researchers noticed several unusual fracking behaviors in the coal. For starters, the
fractures extended much higher than those in conventional
gas fracks, often penetrating several thin coal beds and hundreds of feet of intervening rocks. What's more, over 80
percent of the time the fractures shot out of zone. Even so,
researchers didn't conduct "a systematic assessment of the

extent of the vertical fractures through and above the roof rock shales."

When the unconventional fracking industry arrived on Ruben DeVaughn McMillian's doorstep in the 1980s, industry landmen informed the McMillian family they would be using only sand and water to shatter the coal. Ruben and his daughter Cynthia Ann told the landmen their company could do what it was legally entitled to. But they didn't want to see any wastewater disposal on their land. Nor would they sign a blanket lease agreement. The McMillians also insisted on seeing the labels on any hazardous fluids used during fracking.

As a consequence, the company avoided the McMillian residence and instead fracked around their cross-shaped two-hundred-acre property on Blue Creek in north Tuscaloosa County. One CBM well was drilled only eight hundred feet away from the family's water well. The well had run clean and abundant water for generations, but in 1998 McMillian, a retired BF Goodrich rubber worker, discovered a greasy black gel-like substance coming out of his kitchen tap. Other days, the tap water sparkled with bubbles, as though it had been carbonated. Water from the showerhead delivered a rotten smell, and the well pump house started to rumble and hiss like a pressure cooker. A private report showed that the McMillians' well contained lots of methane.

Elsewhere in the county, Robert Hamrick flushed a toilet and blew his well-house to smithereens. The homeowner suffered severe burns to his arms and face. Other families reported an oily smell in their water or the presence of a slimy substance that reeked of petroleum. Near Lake View, Alabama, rivers of frack fluid from an Amoco well ran down a hill near a house, killing all plant life in its path. The residents lost their well water and later battled cancers of unknown origin. Amoco bulldozed the wayward well in 1991.

The McMillians pressed for investigations. At first, the State Oil and Gas Board and Geological Survey of Alabama hummed and hawed. Cynthia Ann, a high school American Studies teacher, soon discovered why. The agency, like most oil and gas regulators, had a schizophrenic mandate: the Oil and Gas Board permitted water withdrawals for CBM, while the Geological Survey was in charge of not only mapping but also protecting groundwater. Every time the McMillians complained about the changing quality of their water, the board simply ordered the family to purge their well. Amazingly, the water samples the board collected were never tested for methane, nitrogen, or frack fluids, only for normal contaminants such as iron, manganese, and arsenic. Whenever a bureaucrat took an interest in the McMillian case, he or she was quietly reassigned. Cynthia Ann recognized the civic lesson at hand: "We had a cash-strapped state looking for revenue."

When field workers from the Environmental Protection Agency showed up, McMillian showed them an old core well on the property, a legacy of strip-mining in the region. After all the CBM fracking, the well had begun to whoop and howl like a demon. McMillian dumped in a bucket of soap and water, and the flowing gas formed a bubble the size of a man. The curious workers asked McMillian if he had ever tried to light it. McMillian threw them a box of matches, said, "Fire away, boys," and walked back to the house. The sheepish workers later joined him. "Did you light it?" he asked. No, they said. There was too much gas around there.

In 1989, the McMillians hired David Ludder, an environmental lawyer with the Florida-based Legal Environmental

Assistance Foundation (LEAF) to pursue their case. Ludder, now sixty-four, recalls visiting the family shortly after. The McMillians showed him the whistling core well out back. Ludder didn't drink any water from the house well, but he did shower at their home. "I could smell the sulfur," he says. "A couple of days later my skin started to peel off." After hauling water for seven years, the McMillians finally installed a $3,000 filtration and venting system.

The Environmental Protection Agency (EPA) performed tests ten months after the fracking occurred. But as Ludder later discovered, the EPA had done the same kind of bare-bones testing as the Alabama Oil and Gas Board: "They tested for a few 'typical' drinking water quality parameters that would not demonstrate the presence or absence of hydraulic fracturing fluids." That way, the authorities could claim that there was "no documented evidence of contamination."

When Cynthia Ann eventually tracked down the Material Safety Data Sheets on the fracking chemicals—many of them toxic—used in the Black Warrior Basin, she demanded that the board test for those, too. The board balked, and she also got an unusual response from the regulator's chief geologist. Presented with a list of the chemicals, his mouth fell open. "Where did you get that?" he demanded.

Shortly after the McMillians found methane in their water, the Oilfield Review published a revealing article: "Learning to Produce Coalbed Methane." The article celebrated increased methane production but lamented "coal's complex fractures." Controlling "induced fractures" was a problem, it said. Managing the length of horizontal fractures in shallow coal seams seemed impossible, while vertical fractures did not extend as predicted. Moreover, parallel fractures appeared near the wellbore. Many vertical fractures penetrated overlaying zones. Man-made fractures could also "develop offsets at crosscutting natural fractures." But regulators said nothing about these problems to Black Warrior Basin residents.

After his skin-bracing visit, Ludder discovered that the federal Safe Drinking Water Act required state or federal authorities to specifically regulate "underground injection," or what it defined as "the subsurface emplacement of fluids by well injection." In Alabama, the board allowed the injection of gases or water into hydrocarbon formations, to coax more oil and gas out of Class 2 wells. CBM operations also used some of these same wells to dispose of thousands of barrels of wastewater.

In Ludder's view, fracking resembled a newfangled Class 2. It injected thousands of gallons of fluids, many of them toxic, into formations known to contain drinking water. About 20 to 30 percent of the fracking fluids, including diesel fuel and formaldehyde, stayed in the ground, and as a consequence could potentially contaminate groundwater for decades. The majority of the forty fracking chemicals used in Alabama hadn't even been assessed against drinking water standards or for the capacity to cause human cancers, Ludder found. Because hydraulic fracturing looked, acted, and behaved like other forms of fluid injection, Ludder concluded that the technology was an obvious candidate for federal regulation under the Safe Drinking Water Act.

His conclusion was not obvious to the Alabama Oil and Gas Board. In 1989, it argued that hydraulic fracturing didn't constitute an injection because its purpose was to produce methane, and so it was not covered under the rules for Class 2 wells. The Alabama Department of Environmental Management later offered an identical opinion. Neither state agency wanted to take responsibility for protecting groundwater from fracking. "They were actively protecting the industry," recalled Ludder.

84   The lawyer next petitioned the U.S. Environmental Protection Agency to force Alabama to protect drinking water by regulating fracking as a new form of well injection. The EPA denied the petition, declaring in 1995 that it "does not regulate and does not believe it is legally required to regulate the hydraulic fracturing of methane gas production wells." It added that there was no proof of groundwater contamination in Alabama, even though the EPA had never tested any of the chemicals used for fracturing.

The EPA's position ignored its own inconvenient research. In 1987, a lengthy EPA report on the "Management of Wastes from the Exploration, Development, and Production of Crude Oil, Natural Gas, and Geothermal Energy" had detailed a case of contaminated water caused by fracking in 1982 in West Virginia. After fracking a gas well 1,000 feet from the home of James Parson, fluids injected at about 4,000 feet deep by Kaiser Exploration and Mining Company were found in the man's 416-foot-deep water well. A state lab report found "dark and light gelatinous material" as well as methane in the water and declared it undrinkable. Even the American Petroleum Institute, which claims fracking is totally safe and proven, confessed that in this instance there had been a "malfunction of the fracking process." The little-known EPA report confirmed a basic scientific truth: "During the fracturing process fractures can be produced, allowing migration of native brine, fracturing fluid, and hydrocarbons from the oil or gas well to a nearby water well. When this happens, the water well can be permanently damaged and a new well must be drilled or an

alternative source of drinking water found." The report noted
that in scores of other cases, legal settlements sealed by nondisclosure agreements had hidden the scale of the problem. "In
some cases the records of well-publicized damage incidents   85
are almost entirely unavailable for review."

Eventually, the McMillian case ended up in the 11th Circuit
Court of Appeals, where in 1997 the court ruled that fracturing
was indeed a form of underground injection and that Alabama wasn't upholding the law. That decision not only forced
a reluctant EPA to regulate fracturing in the state but rattled
the billion-dollar fracking industry. Halliburton lobbyists were
soon flying to Washington, DC, to keep hydraulic fracturing
exempt from federal laws. In one court briefing, Halliburton
argued that regulation could "have significant adverse effects"
on its business and "would simply make drilling new wells
too costly." In 2002, the Interstate Oil and Gas Compact Commission produced a one-page survey of twenty-eight state
regulators to defend the technology. Even though none of the
regulators had a comprehensive groundwater monitoring program and most couldn't even list the year fracking had begun
in their state, they all placed a check mark in the box indicating
"no record of consequent harm to groundwater." Halliburton
and the industry went on to brandish the unscientific survey
as their best evidence of the technology's safety.

It took more legal wrangling, but the Circuit Court ruling
eventually forced Alabama to rewrite its fracking rules. No
company could now frack without a permit, and the board
outlawed the cracking of coal beds shallower than 300 feet.
Drillers now had to submit detailed plans about depth and
cementing of fracked wells to the board, including a survey of water wells within a quarter-mile radius of the well.

Ludder considered the changes extremely weak, but industry described the new rules as overly strict, unreasonable, and "one of a kind."

The court decision, combined with growing public concern and congressional interest about CBM drilling, forced the EPA to announce a review of the technology in 2000. The agency initially proposed a three-part study on the fracking of coal seams. The first part would review the literature on hydraulic fracturing. A second would conduct field studies, such as measuring methane seepage into groundwater from oil and gas wells. A third and final review would address regulatory options. Thanks to sustained industry lobbying, the study would never progress beyond phase one.

Thirty-five industry groups directly opposed the EPA study. The American Exploration & Production Council argued that no federal oversight of fracking was necessary and that states should be allowed to continue "their effective regulation of hydraulic fracturing." The Independent Petroleum Association of America claimed that "the sheer magnitude of fracturing jobs is indicative that no environmental problem exists that is not already controlled under existing state programs." The presidency of George W. Bush, a Texas oilman, also cast an influential shadow over the proposed EPA study. Bush's vice president, Dick Cheney, CEO of Halliburton from 1995 to 2000, had a direct interest in protecting the technology from federal regulation: his former employer generated a fifth of its revenue from cracking rock. Cheney's 2001 energy policy report not only championed fracking but omitted any concerns about gas migration, groundwater contamination, and out-of-zone fracks. Cheney's paper did note that gas flow rate in certain formations "may be increased by as much as twenty-fold by hydraulic fracturing" and said that most new gas wells in the future would require fracking.

Not surprisingly, the EPA's first study on the technology, "Evaluation of Impacts to Underground Sources of Drink- 87 ing Water by Hydraulic Fracturing of Coalbed Methane Reservoirs," confirmed the views of the Bush administration. Despite extensive evidence of methane migration in Colorado, West Virginia, and Alabama, the agency declared that "the injection of hydraulic fracturing fluids into coalbed methane wells poses little or no threat" to drinking water and "does not justify additional study at this time." The report admitted that the CBM industry had not only, in ten out of eleven coal basins, fracked coal seams containing drinking water but had done so with highly toxic fracking fluids such as diesel fuel. To address such untidy facts, the EPA study saluted a voluntary and unenforceable offer by three major fracking companies to stop using diesel fuel. It left matters there.

The seven-member peer review panel that signed off on the report was anything but impartial: among its members were a petroleum engineer from BP Amoco, a technical adviser for Halliburton Energy Services Group, an engineer with the Gas Technology Institute, an academic who had formerly worked for BP Amoco, and a regulator who had formerly worked for Mobil Exploration. Not one groundwater or contamination expert was included.

A number of prominent critics, however, called the EPA report a sham. Weston Wilson, an environmental engineer employed by the EPA office in Denver, was so incensed that he wrote a letter to Congress, claiming whistleblower protection. Wilson's blunt letter described the report as "scientifically unsound and contrary to the purposes of law." The agency, he

said, had not even "include[d] its most experienced professional staff" in the writing of the report. Moreover, the EPA
didn't conduct field studies to "obtain water quality data near
wells that were hydraulically fractured within or near underground sources of drinking water to determine the extent of
potential risks," though the evidence begged for such science.
Given that hydraulic fracturing was unregulated by either
states or the federal government, added Wilson, the EPA study
had drawn its conclusions of no harm based solely on industry-
submitted data.

Others echoed Wilson's concerns. Representative Henry
Waxman, a member of the House Energy and Commerce
Committee, accused the EPA of making "a faith-based leap
to conclude that injecting toxic chemicals" into the ground
wouldn't compromise groundwater. Geoffrey Thyne, then a
professor at the Colorado School of Mines, argued that exclud-
ing the technology from regulation was "premature, unwise
and goes against the public interest."

As Wilson also pointed out, the EPA report contained much
science that directly contradicted its happy conclusions. For
starters, the report noted that industry had fracked into drinking water near the northern outcrop of the San Juan Basin,
where layers of shale didn't confine coal seams. Gas from water
wells showed the same isotopic fingerprints as gas from the
CBM wells in the basin. Bubbling methane in rivers, exploding
pump houses, and high concentrations of methane in people's
homes indicated "that coalbed methane was following some
conduit from the Fruitland Formation to the surface or to shallow [sources of drinking water]."

The EPA report also admitted that frack fluids could travel,
disperse, and penetrate groundwater, as well as "leak off" into

natural fractures. It referred to such incidents as the "move-ment of fluids outside the capture zone." Fractures rarely behaved as modeled, the report emphasized: "Some portion of the coalbed methane fracturing fluids could be forced along 89 the hydraulically induced fracture to a point beyond the cap-ture zone of the production well. The size of the capture zone will be affected by the regional groundwater gradients, as well as by the drawdown caused by the well. If fracturing fluids have been injected to a point outside of the well's capture zone, they will not be recovered through production pumping and, if mobile, may be available to migrate through an aquifer." Last but not least, the report repeated an enduring truth: accurately predicting the pathway of fractures in coals (or any rock, for that matter) is impossible.

Meanwhile the industry went crazy over slick-water fracks in Texas shales. For years George Mitchell, the CEO of Mitchell Energy, tried to coax methane out of the dense Barnett Shale underneath Fort Worth with no luck. His firm, which was fac-ing bankruptcy, attacked shale rock with injections of carbon dioxide, diesel fuel, nitrogen, you name it. Nothing profitable dribbled out until the company tried a million gallons of highly pressurized water slicked by friction reducers and other chem-icals down a vertical well. Mitchell's fortunes changed and his wells flowed again.

In 2002 Devon Energy bought out Mitchell for $3.5 billion and added horizontal drilling to the equation. The wells went straight down to varying depths and then curved to extend two miles underground parallel to the surface. That way the company could frack more rock over a greater area to harvest more gas in less time. A horizontal well was the equivalent of a bottom trawler with a net scraping everything off the ocean floor. Devon's deviated wells initially extracted seven times more gas than a fracked vertical well. By 2004 the industry was championing slick-water fracks in horizontal wells the way it once embraced Roberts Petroleum Torpedos.

In 2005, President George Bush and Congress ensured that no federal regulations would get in the way. He gave hydraulic fracturing a free pass from the Safe Drinking Water Act. The passage of the Energy Policy Act of 2005 guaranteed that the forthcoming shale gas revolution would also be unencumbered by concerns about groundwater. The term "underground injection," the Act declared, now excluded the injection of natural gas for storage, as well as "the underground injection of fluids or propping agents (other than diesel fuels) pursuant to hydraulic fracturing operations related to oil, gas or geothermal production activities." Industry called the law a "plain language" fix. Critics dubbed it the Halliburton Loophole.

In retrospect, LEAF lawyer David Ludder considers the battle over coalbed methane a trial run for shale gas fracking. The experience taught him that landowners across North America can expect the worst from government and industry. "They can expect to be deceived. They can expect to find nondisclosure about the risks. They can expect denials of responsibility. And they can expect the government to not protect them." Ludder describes the lengthy legal ordeal over CBM as "one of the most disappointing cases. We were right, and we proved that we were right, and the industry lobby convinced Congress to take away what was right." Ludder remains a staunch opponent of fracking, too. "To me the process of fracking can't be made safe, because you can't know the geology. I'm not sure there is a safe way to do it."

Since the EPA issued its fraudulent report, the fracking industry has adopted a new line: it says that its technology never has posed and never will pose a threat to groundwater. "There have been over a million wells hydraulically fractured in the history of the industry, and there is not one, not one, reported case of a freshwater aquifer having ever been contaminated from hydraulic fracturing. Not one," reported Rex W. Tillerson, the chief executive of ExxonMobil, at a U.S. Congress hearing on drilling in 2011.

As for Jessica Ernst, she can no longer count how many times she's heard industry and government officials repeat one of the biggest lies of the twenty-first century.

FIVE

A Revolution
Under Rosebud

JESSICA ERNST PURCHASED her dream property in Rosebud, Alberta, in 1998 for $140,000. The land contained an old homestead, a modern bungalow, a chicken barn, several outbuildings, and fescue grasses. It also had a river running through it. While working in the oil patch, Ernst had driven through almost every small town in the well-drilled province of Alberta, from Trochu to Manyberries. But somehow she had missed Rosebud. The hamlet sat below the open prairie in a deep coulee just an hour's drive northeast of Calgary. Rosebud's residents treasured something many Alberta towns had long since sacrificed for industrial farming or hydrocarbon production: beauty. "I fell in love with it right away," recalls Ernst. "The place oozed charm."

It also oozed history. In the nineteenth century, bison thundered across the rolling landscape as the Blackfoot skirmished with Cree interlopers. The Blackfoot considered the hoodoo-rich geography a sacred or "thin" place, where the distance between heaven and earth collapsed. In the twentieth century, European immigrants from France, Denmark, and Germany saw it as a good place to begin again: they mined the arid country for cattle, grain, or coal.

The newly minted hamlet of Rosebud soon boasted a hotel, a store, a blacksmith shop, a dance hall, a lumberyard, a bank, and a Chinese café and laundry. By the time Ernst arrived, the hamlet had added the popular Rosebud Dinner Theatre and School of the Arts. Barely a hundred teachers, actors, and artists lived in town, but every year the theater attracted more than 30,000 visitors.

From her new home, Ernst could walk out her back door into the Wild West. The Rosebud River, which the Blackfoot called *Akokiniskway* ("the river of many roses"), cut through her property as it meandered through the badlands, an ancient territory where dinosaur bones still poured out of bluffs like junk falling out of a closet. Ernst delighted in the swans and blue herons that fished the waterway. Wild roses bloomed in profusion during the month of June, and cottonwoods, chokecherries and wolf willow dotted the place. Ernst fell so completely in love with the solitude that she soon bought another thirty-five acres of adjoining land. This, she thought, would be a place of healing and a hideaway from the world of men. For the first time in her life, she says, she felt "space" around her.

Lorne Feckley, the land's previous owner, had given Ernst a few pointers about living on the edge of Alberta's badlands. If there was ever an energy crisis, he said, she could grab a bucket and fill it with coal from an outcrop just a short walk away. "You'll always be able to stay warm," Feckley assured her. The water well, located in the chicken barn, had had a clean bill of health since 1986. The Alberta Government record, found by

Ernst, said, "Gas present: No." Feckley did warn Ernst about the blizzards and the loneliness. He didn't think she'd last the year. "It's a hard place," he told her. "You'll never survive here." But Ernst, as was her stubborn way, confounded expectations.   95

After three years of living in Rosebud, Ernst adopted two uncontrollable dogs slated for killing. The Jack Russell and border collie crosses had been abused and locked up for more than a year in a cage. With their black-and-white patches, the siblings looked like abandoned rogues. Bandit wouldn't listen, and Magic was terrified of everything. When the frustrated keepers called Ernst in a last-ditch effort to save the incorrigibles, a close friend advised her to rescue and socialize them.

Ernst wasn't sure she was ready to take on twofold trouble. But she picked up the two dogs after they had been neutered and were awaiting her in the parking lot of a Canadian Tire store in Drumheller. The two scrappers eventually "melted the cement" in her heart, she says. "I learned all about trust and love from Bandit. Magic got that from his brother. We learned together and it was pretty hard work...They had been abused, and they understood me." When not rolling in cow shit or chasing gophers, the dogs were soon accompanying Ernst everywhere.

Around the same time, Ernst picked up a new client for her consulting business. Encana appeared as a new and powerful player in the oil patch in 2002, the result of a merger between the Alberta Energy Company, a firm that landowners disliked due to its aggressive practices, and PanCanadian Energy Corporation, a respected old-style company that specialized in conventional hydrocarbons. With access to more than 17 million acres of oil and gas leases, the new company billed itself as North America's largest independent natural gas producer and gas storage operator. It steamed bitumen out of the tar sands, drilled tight gas in Colorado, sank horizontal wells in British Columbia, fracked tight gas in Wyoming's Jonah Field, and milked the shallow gas machine in the Medicine Hat pool in southern Alberta.

The stock market valued the new energy giant at $30 billion. Gwyn Morgan, a bespectacled mechanical engineer, was Encana's cocky president and CEO. Morgan had a penchant for right-wing politics, climate-change denial, and Buddhism. He also had some bold ideas that, a decade later, would nearly bankrupt Encana. Given the decline in conventional oil and gas plays, along with rising prices, Morgan preached that the industry's future lay in assembling large blocks of real estate containing lower-quality hydrocarbons. By applying hydraulic fracturing and horizontal drilling, Morgan proposed to milk these landscapes with a "military-like deployment of resources." He called the perforation and fracking of large landscapes with an increasing army of wells over several decades "a manufacturing process."

Morgan, who had once adopted the pliable Gumby as a corporate mascot, argued in 2003 that his gas factories were important experiments: "They present huge opportunities to learn and apply technological advancements because the development programs, by their nature, are conducted over many years." Because a gas factory industrialized a landscape, Morgan believed, engaging with the people who lived on that land was "key to minimizing execution risks associated with our developments."

Minimizing execution risks was what Jessica Ernst did for Encana. On large gas pipeline projects such as Tupper and Cutbank Ridge in northern BC, she held meetings, talked to trappers, and informed landowners and the public about routes and right of ways. She also worked on woodland caribou protection plans and made wildlife observations in the field. Her primary job was to make sure Encana consulted with local people, First Nations, and governments, so that everyone understood how wells and pipelines might fragment the forest, increase traffic on the highways, compromise traditional land uses, or affect the fate of endangered species such as bull trout.

By coincidence, Encana tested Jessica Ernst's water well in June 2003. A neighbor had agreed to a conventional gas well that required access from a tiny part of Ernst's land. The neighbor's own water had gone bad after shallow drilling by industry in the past, and he recommended that Ernst request a test. Encana's test described her well-water appearance as "clear," with no fizzing methane. The test found low levels of barium and strontium and no detectable chromium. Her newest client would soon change that. Neither industry nor government informed the good citizens of Rosebud that their community was about to be transformed into a gas factory. Between 2000 and 2002, Encana and Calgary-based MGV secretly drilled several hundred experimental wells to figure out how to coax methane out of 10- to 30-inch-thick coal seams sandwiched in between sands and shales. The coal seams rested 400 to 600 meters (1,300 to 2,000 feet) underground, according to government brochures. In practice, the companies drilled and fracked some pilot wells as shallow as 325 feet. (In 2007, the National Energy Board would reveal that the Horseshoe Canyon coals were so shallow and thin that "some experimentation was necessary in the early wells to identify successful techniques to adopt and unsuccessful techniques to avoid.") The companies were also chasing methane in nearby sand and shale formations. Like the Los Angeles Basin, the prairie around Rosebud had already been punctured like a pincushion for oil and gas. In some cases, the coal frackers used old wellbores to gain access to the shallow coal seams.

The complex geology of the Horseshoe Canyon coals told industry engineers they'd be coping with three unconventional realities. First, they'd have to carpet bomb the landscape with many well sites to get reasonable methane production. Second, well performance would be highly variable. Lastly, the economics would only work based on the aggregate performance of a massive number of wells and compressors. In January 2003, one MGV executive referred to CBM in Oilweek magazine as a "science experiment" that was trying to morph into a commercial development regulators would be comfortable with. An Encana spokesman added that "technological and extraction methods have to be developed to make it economical."

The appearance in the summer of 2003 of an Encana compressor station outside Rosebud hinted that the landscape was about to change dramatically. That same year, the oil and gas traffic abruptly intensified on the highways around Rosebud, along with the number of shallow gas wells dotting the landscape. Yellow lights and flares erased the night sky. Having endured a nasty recent drought and depressed commodity prices, farmers eagerly signed new gas well leases with Encana. But as the density of wells planted on cropland kept increasing, some people started to ask questions about land fragmentation and groundwater. One gas well per section was normal in the province, but now industry wanted to plant four, eight, or even twelve more. The farmers also couldn't tell if the industry was planting shallow gas wells or drilling into coal seams.

Baffled by the commotion, landowners started calling Jessica Ernst for information.

Ernst was as much in the dark as anybody. All she knew was that somebody had planted "an ugly piece-of-shit noisemaker" 870 meters (about half a mile) from her home in a wheat field. Ernst didn't yet know that the coal beds under central Alberta contained so little pressure that industry had to connect scores of wells to a noisy 530 horsepower vacuum cleaner (the size of a two-car garage) to suck out the methane. Nor did she understand that the abrupt appearance of the compressor signaled the beginning of a massive drilling program that would carpet the landscape with thousands of wells. Ernst thought at the time that Horseshoe Canyon was a scenic lookout in the badlands. But to geologists, the 50,000-acre deposit was the perfect testing ground for unconventional gas.

The roar of the compressor invaded Ernst's kitchen and robbed her of sleep. The machine sounded like an airplane taking off. Even the Canadian Society for Unconventional Gas identified the racket of a natural gas compressor booster as an "encroachment" and an "obvious noise source" due "to receptor (residence) concerns" that year. Ernst concluded that some company had installed the noisemaker without any muffling equipment to save money. What she hadn't known yet was that the machine belonged to her newest client: Encana. But Ernst had not moved to her dream property in Rosebud to listen to a piece-of-shit noisemaker all night. She had come for the quiet. With her background as an oil-patch veteran, she fully expected that, with some cajoling, the company and the provincial regulator would fix the problem. When they didn't, Ernst's life went off course in an odyssey that would have tested the patience of Ulysses himself.

By early 2004, Encana's compressor noise had become intolerable. Bandit and Magic paced all night. Ernst had to end her midnight walks because the dogs would dart off in the direction of the compressor and start barking their heads off. Then Encana added a new 800-horsepower compressor to the site, along with several others in the valley, to suck up more methane. Although Ernst and dozens of Rosebud residents flooded the regulator with complaints, little appeared to change. In addition to rising noise and traffic levels, local residents were unnerved by the visits of an Encana land agent. He visited scores of farmers around Rosebud and tried to get them to sign off on a company request to increase the permitted number of shallow gas wells from one to four per section. A signed document would end Encana's requirement to consult with landowners every time it wanted to drill.

To the ears of most landowners, the term "down spacing" didn't sound anywhere near what it meant: a rising density of well sites. Whenever someone asked if the down spacing was to accommodate the fracking of coalbed methane wells, the landman said no. "And even if it was, you'd have nothing to worry about," he told them. To get individual farmers to sign the document, the landman told them their neighbors had already done the deed. But when people bumped into their neighbors at church or the grocery store, they discovered the Encana representative had lied. After a dozen people dropped into Ernst's house in the summer of 2004 with concerns about the document, she thought, "Oh, my God, I better do something."

In late August, Ernst called up Encana's land manager to ask what he was doing. He told her the blanket approval process would make drilling easier and cheaper for the company.

Ernst informed him that she worked for Encana on consultations and that what he was doing was illegal and underhanded.
Guide 56, a provincial rulebook on how to develop oil and gas
wells, clearly stipulated that companies had to consult with 101
landowners about "possible impacts." Moreover, they were
required to answer people's questions. Ernst advised the land
manager that Encana needed to assess and mitigate its cumulative effects. At that point the man laughed and said, "I don't
know what a cumulative effects assessment is, and you don't
either."

Angry by then, Ernst told the man that Encana needed to
hold an open house for local landowners and hamlet residents
to explain what it was doing. The manager replied that the
company wouldn't have to hold an open house if everyone
signed the letter. He also hinted that if Ernst wanted to keep
working for Encana, she should keep her mouth shut. That
was the wrong thing to say to Jessica Ernst. By the end of their
conversation, the manager had reluctantly agreed to organize
a community open house. But he reneged on his promise and
Encana's landman continued to pester landowners to sign the
blanket approval document. "Holy shit," thought Ernst. "Not
only am I being lied to, but my community is being lied to, too."

Ernst had quit a project a few years earlier for ethical
reasons. Predator Energy ("What a fucking name for an oil
company," she points out) had hired her to do a pipeline application during the spectacular Ladyfern gas play in northern
BC. Several competing gas companies were trying to plant
straws in a massive but conventional methane milkshake buried deep in muskeg country. In their haste and greed, the firms
had sucked so hard on the gas field that they flooded it with
water and destroyed it. Ernst became an unwitting witness to

a common crime in the patch: she watched as a Predator executive broke into a well shack to obtain valuable production data owned by Murphy Oil, a competitor. The theft became the subject of a major lawsuit. "I quit because Predator broke the law," Ernst says today. "I witnessed them and took photos of them trespassing. I couldn't believe what I was seeing."

Her conversation with Encana's land manager left Ernst in a quandary. A major client had broken the law and was misleading her community. She went for a walk with Bandit and Magic on the prairie and she talked over the situation with them. She explained to her black-and-white companions that Encana paid the bills for her staff (at that time she employed numerous contractors) and that resigning might have serious repercussions. "This might harm us," she told them. Ernst cherished her dogs and their unqualified love for her. Bandit, she recalls, gave her what she describes as an "I'll die for this" look. Ernst knew that her resignation might change her life and could possibly lead her "down a road to hell." She had no idea.

Her resignation letter, sent on September 9, 2004, was pure Ernst. It informed Encana that the company's "lack of ethics and current underhanded consultation practices" in Rosebud violated provincial regulations. Encana's actions, Ernst said, made a mockery of the company's own policies about acting ethically and "conducting public consultation" in accordance with the rules. Ernst added that she couldn't understand why Encana had hired Ernst Environmental Services to do detailed public consultation for a few kilometers of pipelines in British Columbia but couldn't be bothered to do anything about the undeclared yet massive resource play around Rosebud. "There are thousands of wells and kilometres of pipelines and roads proposed without adequate or diligent consultation or warning or mitigation planned... I cannot in good conscience work for a company operating in such a manner."

Ernst's letter got two immediate replies. An anonymous email from someone identifying himself as an Encana employee noted that Ernst had "sent a warning shot across the bow of the evil empire, creating a huge wave." The other email came from Mark Taylor, manager of Encana's Wheatland Business Unit, thanking Ernst for her comments. "Encana has not presented enough clarity regarding its proposal to you and the other Rosebud community members," admitted Taylor. He promised an emergency meeting, held the following Sunday.

At that meeting, attended by hundreds of local residents, Encana executives said there would be no CBM wells at Rosebud, because all the natural gas had vented out of the coals years earlier through outcrops. The company also promised swift action on Encana's speeding drilling trucks, including a traffic survey. Incredibly, the resulting survey found that 40 percent of the traffic running through Rosebud was oil- and gas-related, and nearly 80 percent of the seventy-six vehicles recorded on the first day of the survey were caught speeding. Shortly afterward, Encana fired the land manager who had touted the blanket approval process. Another energy firm hired the man the next day.

Later in September, Stacy Knull, an Encana vice president, promised Ernst more openness as well as action on the noisy compressor. Knull, who today runs his own unconventional oil and gas company, invited Ernst to talk to a group of Encana employees in downtown Calgary about the importance of full disclosure. Everyone expected that the subject of Ernst's talk would be something Knull billed as "Surface Land Brainstorming and Optimization." Instead, Ernst stunned the group by talking about lying and rape. She didn't mince words. She first asked the group what they would do if they caught one of their children lying. Many answered that lying doesn't work and that the truth always comes out.

The Calgary group fell silent when Ernst asked them, "How many of you know someone who has been raped?" Five people raised their hands. Rape, she coolly explained, has nothing to do with sex but everything to do with control and power. "A landman comes to a farmer," Ernst told the group, "and says Encana will give you a lovely little kiss, some money, and one well. But the landman doesn't tell the farmer that he is about to be raped"—in this case with compressors, traffic, more wells, and lies about so-called responsible development. When a resource company approaches landowners or a community like Rosebud without full disclosure of the facts, the stakeholder "becomes a victim of rape," said Ernst. The only way to remove such a stain is through full and open dialogue, she added. Two Encana employees later told Ernst that they found her shocking analogies "excellent, eye-opening, and appropriate."

As Rosebud bubbled with troubled talk about resource plays and fracking, Canada's fifth largest bank, CIBC, circulated a glowing investors' report about coal's unconventional virtues. The report, entitled "Moving Down the Resource Triangle," contained lots of facts that neither industry nor government had ever shared with citizens of Rosebud, or with other Albertans, for that matter. The document began by noting that "all sources of unconventional gas will be necessary to meet long term demand growth." It described the pace of growth in the Horseshoe Canyon as "astounding" and portrayed CBM as a capital- and energy-intensive resource best exploited by "assembling a large land base and drilling large amounts of wells at a time." The report predicted that companies would drill 2,500 to 3,500 wells in 2005, and nearly 4,500 every year by the end of the decade. "It is imperative," added the report, "that CBM operators spend appropriate amounts of time educating and consulting landowners and other stake holders prior to development."

In her own research, Ernst found a similar report by Canada's National Energy Board. The NEB report said that the continent was running out of conventional natural gas and that Canada was at a crossroads. It described CBM developments as an uncertain resource, due to concerns about well density and water impacts, and it noted that the Horseshoe Canyon play might require 50,000 wells over a twenty- or thirty-year period. Portraying the unconventional play as a multi-well "manufacturing process," the report suggested "a blanket approach" for drilling programs instead of the usual one-well-at-a-time application system. That way, industry could move faster. The report didn't contain a word about cumulative impacts. Nor did it question the practice of drilling more wells over greater expanses of land to release smaller quantities of gas. When an industry friend of Ernst's read the report, he shook his head. He told Ernst it was going to be "a long arduous process to ensure that CBM development is managed properly for minimum cumulative impact. I am not even sure that the term minimum impact can be used for CBM because I personally see no practical way to do it."

If the Alberta government or Encana had asked Jessica Ernst to prepare a cumulative effects assessment on how a fifteen- or twenty-year-long 50,000-well coalbed methane development might industrialize the rolling farmland of central Alberta, the forthright consultant would surely have done so. "First of all," Ernst says, "you have to tell the people what you are doing, because you are putting groundwater at risk, a vital resource used by the entire community. So you show everyone the plans and the map."

Next, says Ernst, you explain to landowners that the resource is difficult to extract and will require two or three times more wells planted in the ground than conventional gas does. You inform people that some coal seams are under such low pressures that industry will have to plant scores of compressors on the prairie to vacuum up the methane. "You show where you are going to drill and map the geological formations," Ernst continues. "You explain where the groundwater is and that 300,000 rural residents depend on the resource. You also explain the coal seams often contain fresh water and that industry will be drilling very close to groundwater zones with no buffer. You don't say you'll never perforate and fracture coal seams anywhere near aquifers, because accidents happen. If you lie to people, they'll get in an uproar."

Following all that, says Ernst, you explain the reason for the massive industrial gas boom. "The situation is dire. We are running out of natural gas in North America. The price is going up, and we need more gas production now. We have obligations under NAFTA, and this is what we need to do. Then you come up with a protection and mitigation plan for landowners. Because the project is so high risk, the government must first drill community groundwater monitoring wells and gather data for two years before any company starts to drill and frack in the region. That way you've captured seasonal fluctuations in groundwater chemistry. And that way you can answer questions about changes in water quality. How much methane is already in the water? How much is biogenic? Is there ethane in the water? Is there evidence of previous oil and gas activity, such as kerosene and diesel fuel? During a drought, does the methane increase in groundwater? Once you have that data in hand and report it publicly twice a year, then you can start drilling and fracking—in one area first. And if something goes wrong with the water, you can identify the source and prepare a plan. Do you move the entire community? Do you leave the community there? We know venting gas is not safe. Or do you leave the community there and pipe in water from the Red Deer River?" Ernst knew just what Encana needed to do in the area if they wanted to pursue CBM development legally and responsibly. But nobody in government wanted to listen to Ernst.

Although an Alberta government–formed Multi-Stakeholder Advisory Committee met regularly throughout 2004 to study coalbed methane, participants later called it a calculated sham. As one rancher recalled, "The entire government has only one top priority, and that is to turn our hydrocarbon resources into cash as quickly as possible." An industry heavy oil geologist who took part in the committee's water task force later described it as completely biased. Nor was the committee's report taken seriously by government. Two years after the stakeholder group tabled its recommendations on best practices, a legal journal, *Environment Law*, would report that "there has been little in the way of policy or guideline implementation."

On October 21, 2004, Encana finally held an open house to talk about its coalbed methane plans for the area and to answer complaints about what local residents called "inappropriate consultations." A TV reporter videotaped the meeting while the company offered refreshments and beef on a bun. Mark Taylor told the crowd that Encana would never frack coals near the community's water aquifers and that the company would take great care when drilling gas wells on uplands above the valley floor, where many people had sunk their water wells. He also announced a $150,000 grant to the Rosebud Theatre and plans for a second noise study. At the beginning of the meeting, the company handed out an aerial map of the proposed noise survey study area. Of several private properties on the map, only Jessica Ernst's residence was identified by name. That blatant violation of her privacy outraged Ernst. An industry executive laughed as he gave her a copy of the map. "You'll be getting some angry visitors very soon," he said. And she did.

SIX

Criminal Threats

IDENTIFYING ERNST'S COORDINATES on the map was no accident. The consultant had proven that the first noise study was fraudulent, and her boldness still rankled the company. Encana quietly began that ill-advised study only after Ernst and her neighbors had lodged more than thirty complaints about the company's compressors. The secret survey appeared to have two goals: silence the complainers and provide the regulator with a report that suggested there was nothing wrong and that Rosebud was a naturally noisy place anyway. Ernst couldn't help but notice a raft of irregularities in the study. For starters, as the noise experts wrote in their report, the company turned off its noisiest compressor. Not once during the study period did the loudest compressor shake, rattle, and roar like it had for the past year. The noise testers also placed their microphone nearly one kilometer (over half a mile) away from where it belonged. Ernst and other landowners called the study dishonest and demanded that the company fix the problem. Instead, Encana promised another study run by the same experts. According to Encana vice president Stacy Knull, the goal of the second survey was to help experts find the best way to quiet the noisemaker.

Suspecting that Encana was too cheap to fix the problem, Ernst turned to the province's energy regulator for help. The Energy and Utilities Board had been regulating oil and gas activity in Alberta since the 1930s. Like most workers in the oil patch, Ernst considered the organization, which oversaw some 300,000 wells and associated pipelines, among the best in the world. Furthermore, the EUB had a mission to develop energy resources fairly in the public interest. "I thought that EUB was world-class and that they would do something about the noise," Ernst recalls. When the board accepted Encana's first noise survey as legitimate, Ernst thought maybe the regulator didn't understand its own noise directives. "I'll educate them," she decided.

The second sound survey, which happened just before Christmas, wasn't much of an improvement. Fearing more "funny business," Ernst kept a detailed noise log. Sure enough, the company operated its offensive noisemaker much more quietly for the duration of the survey. The study also violated regulatory guidelines for noise testing by moving the microphones thirty meters away from Ernst's house, after it had been set up initially at the legal fifteen meters. Ernst asked a retired noise expert to come to her place to inspect what was happening. He chuckled as he walked to and fro and confirmed that both studies had been fudged, to make it look as if a rule-breaking company was actually compliant. The EUB, which ignored raw data showing that noise levels at Ernst's house violated the law for 67 percent of the study nights, later approved the second study.

Ernst was furious. She sent the board numerous letters detailing the "reporting inaccuracies" in both studies. Her damning critiques sparked an uproar at the EUB. The Calgary executive of a noise-measuring company later told Ernst that the regulator had hauled in the province's noise-measuring experts and given them a tongue-lashing. "Don't do this bullshit again and get caught," they were told.

Dumbfounded by the board's willingness to allow such rule-breaking, Ernst spent some time at her small cabin in Saskatchewan. The cabin offered her what she had by now lost in Rosebud: peace and quiet. But upon her return to "CBM land" in January 2005, Ernst was greeted by a yellow scourge on the horizon. The pollution made by the flare stacks and compressor stations left a metallic taste in her mouth, and Encana's "high-pitched compressor noise" once again invaded her home. She called Encana's emergency number to report another disturbance. In her absence, nothing had changed.

That same month, Rosebud's water reservoir blew up. The local *Strathmore Standard* reported that an "accumulation of gases" had caused the explosion, which badly injured the county's water operator. The operator had been trying to thaw a frozen inlet pipe with a propane torch when a blast damaged the entire reservoir.

A geologist friend of Ernst's read the story with some alarm. He wrote to her that it "sounded like there could be absolved methane coming out of the drinking water." Methane, he explained, "has a reasonably high solubility in water." It had no odor and was extremely dangerous. "The contamination could be natural but it could also be related to industry activity in the region," the geologist added. Ernst dismissed his concern at first. Everyone in Rosebud, she said, had been told that the operator "had lit a propane torch to thaw something, left it there, come back, torch had gone out, and he didn't think and relit it." The propane, however, loaded with mercaptan, would have left a strong warning smell in the air yet a local paper reported that the operator "had done his checks," and "was unable to detect the gases by smell." That spring a visitor to Ernst's house couldn't believe the noise. "Is it a train?" she asked. "No," snapped Ernst. "That is Encana." As the compressors droned on at high volume, Ernst fired off more letters. One went to Florence Murphy, Encana VP Community Relations. "I request please, for the sake of my health and that of others and the environment that Encana turn the compressors off at night until the noise problem is fixed. That will ensure a speedy remedy takes place and I'll get my sleep back." Good neighbors, which Encana claimed to be, did such things, Ernst added. She also sent along a World Health Organization report showing that noise between 37 and 42 decibels not only annoyed the hell out of people but resulted in sleep disturbances, hearing loss, and adverse social behavior. The compressor noise had reached 51 decibels outside her home. Her letter also cited the Noise Pollution Clearinghouse: "Noise polluters are like bullies in a school yard. They are basically saying, I don't care about you and the effect my noise has on you. It's a power issue."

Ernst didn't know it at the time, but she wasn't the only landowner in the west who was fed up with unconventional compressor noises. In 2001, after enduring the incessant whine of a compressor station for two years, Dave Bullach, a welder in Gillette, Wyoming, had "stormed out of his house at midnight...with a rifle and shot at the compressor until a sheriff's deputy hauled him off to jail," the *New York Times* reported a year later.

Ernst decided to report the compressor to the Drumheller RCMP, describing it as "an Encana disturbance of the peace." Her novel complaint worked wonders. Silence abruptly returned as Encana shut down one of its two compressors during the evenings "to address concerns expressed by residents in the Rosebud area." Ernst wrote in an email to friends that it "was pure bliss to have my house back after so much incessant Encana whining inside my home." But it was a temporary respite.

In April 2005, the *Globe and Mail's* business magazine published a story called "Life Inside a Science Project," on the coalbed methane uproar in central Alberta. Comments by Ernst and several Rosebud citizens on the lack of consultation figured prominently. Other experts questioned the long-term economics of fracking coals. In response, Gerard Protti, Encana's executive vice president, wrote a letter to the editor, saying the article "fell short of what Encana expects from a business publication."

After the *Globe* article ran, more and more local farmers started to drop by Ernst's place with reports of water problems. Some wells had lost volume, and others had gone murky. Chris Gerritsen, a nearby water-well driller, alerted Ernst to severe problems just west of the hamlet in the Rosebud Valley. Encana might have fracked into an aquifer there—several wells had gone bad. The driller explained that the company was fracking with nitrogen and other chemicals, and the result was a mess. The news was alarming. If Encana was using "nitrogen and other substances" in their fracking and refracking of CBM wells in the Horseshoe Canyon, Ernst wondered, what potential poisons might humans and livestock be drinking?

Ernst had noticed some changes in her own water, too. A mysterious white fog issued off it. The household kitchen taps whistled loudly, like a train. Both Bandit and Magic turned their noses up at the water. Only later would she realize that losing your groundwater is like going bankrupt. Debt happened, as Hemingway wrote in *The Sun Also Rises*, in two ways: "gradually and then suddenly." For now, given her hectic work schedules—she flew north often to Fort St. John, Fort McMurray, and Edmonton—Ernst remained focused on the noise. The deafening compressors had started up again, assaulting her entire being.

News about the turmoil in Rosebud had spread, and several oil and gas companies queried Ernst about how to do things differently. She advised them to quiet their compressors with special sound suppressors and then house them in farmlike buildings instead of erecting naked noisemakers on the prairie. Pioneer Resources did just that. Its CEO wrote to Ernst directly, saying he was sorry to hear "that your Encana situation has not improved and it has degraded your quality of life. We hear through Encana people around the Acme area that they are being blocked for every permit they seek and it is for obvious reasons—the total lack of respect for anybody but themselves. Pioneer has been bullied by Encana in Western Canada and even in the Gulf of Mexico. It is sad to see a company that could be a stellar role model behave so abusively."

Ernst hired Edmonton lawyer Richard Secord to dash off a letter to the Energy and Utilities Board itemizing her growing concerns about rapid CBM development around Rosebud. The letter methodically cited a long list of impacts, including water taps that wheezed and sang, dangerous levels of truck traffic, roads ripped up by Encana's drilling and fracking convoys, and residents' loss of privacy and solitude. "Our client," wrote Secord, "is of the view that CBM remains an unknown at this time.

In order to protect Alberta and its citizens, it is imperative that
CBM is regulated responsibly, slowly and with caution."

The board never replied to the letter. Nevertheless, Ernst
badgered the regulator throughout the summer of 2005
about the bogus noise studies. She shared much of the correspondence on a BC CBM Listserv and in emails which went
out to Rosebud residents, local politicians, and several hundred landowners across North America. In August, she told
the EUB that its proposal to do another so-called blind study
without notifying Encana, so the company couldn't turn off
its compressors, was stupid. Why order another sound survey
until "the past non-compliant matters have been satisfacto-
rily resolved?" asked Ernst. The board might have the most
stringent noise-control regulations in the world, she added,
but "what good is it if the EUB does not regulate proponents
who do not adhere to it?" Furthermore, she argued, Encana
knew that its compressors violated the law: "Why else do a
noise study with the noisiest compressor turned off?" As an
ordinary citizen, she had once lived a quiet and private life in
Rosebud, she noted—"I was not expecting to have my legal
right to quiet enjoyment of my property violated by Encana.
I was not expecting to have to fight so hard and long and still
not have my quiet back. I was certainly not expecting the EUB
to enable a company in its cycle of abuse." She also suggested
that the EUB enroll its staff in "ethics and integrity training." A
sympathetic EUB staffer replied, warning Ernst that Encana
was so powerful in the province that "it is unlikely the EUB's
noise staff will do anything."

Near the end of the summer, Encana invited Rosebud residents to a day of golf and theater. The occasion, the invitation
said, would give "Rosebud residents and Encana employees" a

chance "to mix, mingle and get to know each other in a relaxing and casual atmosphere." After a game on the green, the minglers would watch a Rosebud Theatre production of The

Village of Idiots. To Ernst, the whole thing sounded like a corporate PR ploy to "come on down, play golf with the devil and shut up." Another three CBM wells had been drilled north of her property. Given the incessant noise from the drilling and the compressors, Ernst replied to Encana's invitation tartly. Why, she demanded, did the company want to "mingle and play with me" when it hadn't yet solved "the compressor noise impacts in my home"? Instead of buying the community theater tickets, she suggested, it would be more useful for the company to spend money "to adequately plan and mitigate the negative cumulative effects of CBM fracking before they happen."

By now, Ernst's water taps ran black bits of coal. The water felt slick to the touch. Ernst often rubbed her thumb and finger together as she washed dishes, wondering, "Why is my water slippery?" Lightly steamed vegetables and boiled pasta seemed to disintegrate. An odd pink bacteria-like slime had also appeared in her toilet.

On September 9, 2005, the compressors woke up Ernst so violently at 4 a.m. that she grabbed the phone. She called Stacy Knull, Encana's vice president for the region, waking both Knull and his wife. She gave him an earful: you can't use bad science and fudge data, she said, to repair a noise problem that could have been avoided with proper attenuation in the first place. The two talked about solutions. Ernst suggested moving the compressors. Knull didn't think that was a good idea. Ernst asked how she could ethically sell her place with so many "shit noisemakers" in the neighborhood. At that point, Knull offered to buy her property: "We will buy your place, so that the noise problem becomes ours... Do you understand?"

The buyout offer caught Ernst by surprise. She told Knull she didn't think it was right for her to have to change her life just because of Encana's bad planning. Her house had stood along the river for twenty-five years before the compressors arrived. "The problem is Encana's and only Encana's," she insisted. But she promised to give Knull's proposition serious thought.

Shortly afterward, Ernst's heavy oil geologist friend and his wife visited for dinner. At the kitchen table, the geologist immediately noticed the agitated state of the water in Ernst's drinking glasses. The couple had sipped Ernst's well water many times before and found it beautiful and clear. Now it was acting like ginger ale and spitting out of the glasses. The geologist said he had never seen such crazy effervescence. Knowing that groundwater doesn't change that dramatically unless something catastrophic has happened in the aquifer, he urged Ernst to get her water tested right away. The fog coming out of her taps might be CO_2, the geologist added, but he wasn't sure.

After that evening, Ernst and her nephew, Derek, started to do some serious Internet reading about gases and water. The nineteen-year-old, who loved to wander along the Rosebud River looking for frogs and muskrats, often visited his aunt on the weekend. There were problems at home, and Rosebud had become Derek's refuge. After his visits now, though, Derek's eyes ran and itched for a few days. He wondered if his eye problems had something to do with his auntie's well water. Ernst's eyes were also chronically irritated when she was at home, she realized. Her skin turned bright red after a shower or bath, but she had attributed that blotching to menopause.

Through a Google search, aunt and nephew came across a 1982 pamphlet called "Methane in Water Wells," put out by the Michigan Department of Public Health. Methane is lighter than air, they learned. It will mix with water at low temperatures but gas off at temperatures above 42 degrees Fahrenheit. Water wells underneath the Antrim and Coldwater shales in the northern part of Michigan sometimes collected small amounts of nitrogen and methane, said the pamphlet. (Despite much opposition, Encana would later frack these formations, too.) Testing for methane was simple, said the flyer. All you needed was "a plastic, narrow-mouthed milk carton and a book of matches."

For a lark one evening, the two became experimenters and gave it a try. Ernst filled a plastic jug with well water, then placed her hand over the mouth of the bottle for a few minutes. (When she learned later that methane molecules are small enough to pass through the skin, she stopped using her hand as a cap.) The pamphlet had promised that "the presence of methane will result in a brief wisp of blue or yellow flame." But that's not what Ernst and her nephew beheld. "The bottle took off like a rocket in my hands," recalls Ernst. Half of the jug melted. The two looked at each other in disbelief. Derek later flew into an uncharacteristic rage. "Auntie, this is our water. This is our place!"

In between work assignments, Ernst made more inquiries about testing for methane in water. She also asked Encana what the terms for a buyout would be. A company lawyer replied that three appraisers (two chosen by Encana and one by Ernst) would view her property and that the company would pay the average of the three estimates. It would also pay for her to move. In return, the lawyer told her, "You sign a release of Encana and all of its affiliates from and against any and all claims, actions, causes of action, damages, losses (including but not limited to loss of income and loss of business) and expenses." Given the fiery state of her groundwater, a public resource, Ernst decided to ask for more information first. Before signing anything, she wanted a list of fracking chemicals used by the company, the names of other landowners who had filed water complaints, and any reports completed for Encana on problem gas wells within a five-kilometer radius of her property.

OF THE MANY calls and inquiries from landowners Ernst was fielding with wearisome regularity, one had come from Laura Amos, an outfitter in Silt, Colorado, who had read the *Globe* article. Through emails and a phone conversation in early October, Ernst pieced together Amos's story. The forty-two-year-old mother lived with her family in what she called "Encana's Industrial Wasteland."

Amos's ordeal had begun in 2001 when Encana invaded Colorado and bought out Ballard Petroleum to exploit tight gas formations. In the process, the company took over responsibility for a Ballard gas well, drilled and fracked just a thousand feet from the Amos home in Garfield County. When the frack job went out of zone, it blew out the Amos water well and turned it into a geyser of mud and air. The water turned gray, bubbled like pop and smelled horrible. Tests showed it contained 14 milligrams per liter of methane.

The Colorado Oil and Gas Conservation Commission, whose mission is both to protect public health and to foster oil and gas development, called the gas "transient" methane. The commission fined Encana but told Amos not to worry, just to be sure her family carefully ventilated their cabin. The regulator assured Amos that methane was safe, adding that humans produced it naturally. Encana agreed to deliver water for three months. When the fizzing and the smell in their water well subsided, the Amos family assumed the aquifer had cleared and started to use their water for drinking and bathing again.

In 2003, doctors diagnosed Amos with a rare adrenal gland tumor. After surgeons had removed the tumor and the gland, Amos came across, on the Internet, a 2002 memo written by Theo Colborn, a world-renowned expert on how chemicals affect the human endocrine system. Colburn expressed concerns about the use of a new fracking fluid chemical called 2-butoxyethanol (2-BE) in gas wells in Colorado's Grand Mesa National Forest, since studies showed that the odorless chemical caused adrenal cancer in rats and could raise hell in domestic animals if it got into groundwater. Amos started asking more questions and writing letters. She became, as she put it at the time, "one mad mother."

Encana later admitted that it had experimentally fracked with 2-BE by Amos's home for nearly a month. At the time, however, the company denied any wrongdoing and suggested that Amos's exposure to the chemical probably came from household cleaning products. Meanwhile, a new investigation of her family's water well found higher levels of methane. The Colorado Oil and Gas Conservation Commission reckoned it came from the formation Encana was now fracking. The regulator served the company another notice of violation, for which it later fined the firm $94,000. The company paid but denied any responsibility. Amos discovered that, of thirty-four violations issued by the Colorado regulator between 1997 and 2004, Encana accounted for 71 percent, even though the firm had been operating in the state for only three years. One 2004 fine was a record $340,000 for the company's "failure to prevent the contamination of fresh water by gas" in Mamm Creek.

Laura Amos told Jessica Ernst about a landowners' conference being organized by the Oil and Gas Accountability Project. It would take place in late October in Farmington, New Mexico, and the subject was "Toxics in Our Communities." Amos was going to present, and she asked if Ernst wanted to come. Ernst wasn't sure she'd have the time but eventually relented.

During the conference, Amos and Ernst chatted for hours, comparing notes about the all-consuming nature of gas development and its disruptive effects. At one point, Amos confided that she was thinking of settling with Encana out of court. The whole awful battle had worn her down, and even state regulators portrayed her as a crazy woman, she told Ernst. The company had offered the family a generous deal, though it came with a serious gag order. If the Amos family signed a confidentiality agreement that prevented their heirs from ever speaking about groundwater contamination and fracking, Encana would buy their home and help resettle them with a million-dollar cheque.

Ernst says that her stomach fell out of her body at the revelation. She asked Amos, "How could you ever sign anything that denies your daughter freedom of speech for money?" Laura's husband, Larry, Ernst recalls, said something about the need to get out of a sacrifice zone and move on to safety.

After Ernst returned to Rosebud, she wrote Laura Amos with some news: "I came home to our energy regulator increasing allowable noise levels for industry. I am sure this is in direct response to my evidence of Encana being non-compliant. I feel so mad I could spit! And I feel so sick about this. Every time Encana is found to be non compliant, the regulator changes the rules to fit Encana's noncompliance. How gross is that?" But Amos didn't reply. The outfitter had decided to negotiate a buyout with Encana and didn't want to jeopardize the deal. Doug Hock, an Encana spokesman, later explained the nature of the arrangement with the Amos family to the *Denver Business Journal*: "It was clear to us that we hadn't impacted her well—and we received a fine from the COGCC, which we also believed was not valid—but nonetheless we decided to move on with the issue at that point." Hock outlined the company's reasoning. "At the end of the day, it's a decision of—do you want to go through litigation and press attention? Or just settle it and move on? It's a business decision, and businesses make those decisions all the time."

On November 2, Ernst collected water samples and drove them to a Calgary lab to be tested for dissolved methane. She included samples from a neighbor's farm and one from a bed and breakfast in town. Fiona Lauridsen, who lived with her family on land just west of Ernst's place, had noticed problems with their water, too. Around the same time, Ernst got an interesting note from local water-well driller Chris Gerritsen, who confessed that he had fracked water wells in the past to enhance water production but did so no longer. "In my experience," he wrote, "fracturing… is a dangerous game to play with nature. I had found that there is very little control over what can get broken, altered, shifted and squeezed and it is not reversible. Although we have improved some wells, the risk of damage to the surrounding aquifers and other people's wells was extremely high. I quit this practice years ago when I realized nature in this state should not be messed with. Fracturing fluid or gas will always move to the weakest point which is what earthquakes are all about—fault lines and shifting. Scary stuff especially when it cannot be fixed...As much as the engineers think they have fracturing figured out, they are getting more dangerous weakening the earth, thus creating a problem that cannot be repaired. I don't see a backup plan if our water is jeopardized." Gerritsen promised to drop off an industry report that shed some light on water problems in the region.

123

In early November, the EUB sent Ernst a draft copy of its new noise-control directive. The directive had been rewritten to accommodate CBM development and industry's growing need to plant more compressors. After reading the lengthy document, Ernst dashed off a warning to her frac news list and the BC CBM Listserv. She alerted landowners that the regulator was going to allow industry a 5-decibel increase during the winter months, "even though noise usually gets louder as it gets colder." Although the regulator reasoned that there was more demand for gas in winter and therefore more compressors would be needed, "I think their reasoning is that rural people do not need to sleep in winter," wrote Ernst sarcastically. Instead of punishing Encana for breaking noise levels of 50 decibels, the board had changed the rules to make Encana's actions legal. "Rural Albertans will soon become Mad Hatters in the name of public over-consumption...oops...interest," she wrote. Ernst ended her letter with this fateful line: "Someone said to me the other day: 'You know, I am beginning to think the only way is the Wiebo Way.'"

Everybody in Alberta knew about Wiebo Ludwig. The evangelical Christian and his large extended family started a full-scale battle with the oil and gas industry in 1996. After plumes of sour gas, a deadly neurotoxin, trespassed on Ludwig's rural property, killing livestock and sickening members of his family, Ludwig demanded that government regulators do something. When, after five years of respectful pleading, they failed to answer his concerns, Ludwig, the son of a Dutch Resistance fighter, declared war. It started with nails left on the road and graduated to deliberately punctured tires. Next came downed trees on oil service roads. Then came the unrelenting monkey-wrenching of well sites and pipelines. Between 1996 and 1998, hundreds of remote facilities were attacked and disabled under the cover of darkness. The violence escalated to shootings, death threats, and bombings. During the campaign, a mysterious group of saboteurs (mostly members of Ludwig's immediate family) destroyed more than $10 million worth of industry property, much of it owned by Encana's precursor, the Alberta Energy Company. Ludwig also dumped sour crude in the offices of the provincial regulator to demonstrate the offensive nature of the product.

The RCMP spent more than a million dollars on a lengthy undercover investigation of the mayhem, breaking the law themselves in the process. In 2000, the courts found Ludwig guilty of bombing a Suncor well site, encasing a Norcen Energy well in concrete, and counseling an RCMP informant to possess explosives. After his release from prison, Ludwig encouraged his religious community to build windmills and straw-bale homes and to use plant-based diesel fuels. The regulator regarded the affair with Ludwig as one of its greatest embarrassments because it failed to uphold its own laws.

Ernst posted a thorough critique of the EUB's new fifty-two-page noise directive to her frac news list and the BC CBM Listserv, highlighting in color the most offensive parts. She also eviscerated the board's bureaucratic jargon. To the board's declaration that changes had been made to "ensure the application of technically viable noise assessment methodologies," Ernst added in bold: "If science is manipulated and misused, it does not matter how technically viable the methodologies are." 125 The draft directive said the board wanted to "provide accountability through a meaningful compliance assurance program." "What does this mean?" asked Ernst. "I provided a number of reports of clearly evident non-compliance by the proponent impacting my property and the EUB dismissed them."

The scathing critique continued. Where the board recommended best practices, Ernst wrote in bold, "Rather than state a best practices approach is required, EUB needs to state that all proponents must adhere to the noise control direction without exception." The board claimed it sought balance "by considering the interests of both the nearby residents and the facility owner/licensee." Ernst responded, "There is no point of compromise or balance in human tolerance of noise." And on it went.

Ernst followed up her postings by dashing off to the Yukon on a week-long tour to give public talks in Whitehorse and First Nations communities on the impacts of fracking. Two oil and gas companies had expressed interest in fracking shales and coals in the northern territory, and the Canadian Parks and Wilderness Society wanted a frontline expert to comment. Ernst explained to Yukoners that unconventional resources made unconventional demands on rural communities, including hideous traffic and noisy compressors: "I've worked on a variety of different developments in agricultural areas as well as Crown land and mountainous areas, and I have never seen such an industrialization of the landscape." She said the fracking in Rosebud had been experimental and the companies, including Encana, hadn't disclosed the chemicals they used. "We have asked. Because it stands to potentially impact aquifers, we think it's an important thing for communities to know what the fracturing chemicals are. The regulators won't disclose it either."

She arrived home from the Yukon in early December to find in her mailbox a three-paragraph letter from Jim Reid, manager of compliance and operations at the EUB. Reid accused Ernst of attacking the regulator by undertaking "an intensive writing campaign as a means to pressure the Alberta Energy and Utilities Board (EUB) to rule that Encana has not met the regulator requirements for noise control in the Rosebud region." Reid pressed on to his main point. "Even though two previous surveys conducted by a reputable acoustical engineering firm were technically defensible and did demonstrate that Encana was compliant," he wrote, the board had offered Ernst a separate "blind" noise survey at her residence. "Rather than accept this offer you have chosen to perpetuate accusations that the EUB has not been responsive to your concerns." Reid's letter then accused Ernst of circulating "internet untruths that the EUB has unilaterally made significant changes to the Directive."

At the end of his letter, Reid banned Ernst from contact with the EUB. "What I cannot and will not accept," Reid wrote, "is your threat, veiled as something said to you as a means to incite people to resort to the Wiebo Way. Criminal threats will not be tolerated. We are deciding on how best to work with the office of the Attorney General of Alberta and the RCMP to register our concern and ensure the protection of the public including our staff." Until security issues had been resolved, Reid said, he had instructed his staff "to avoid any further contact with you."

Ernst read and reread Reid's words in disbelief. She had expected the board's letter to contain an apology and a promise of action on the noise infractions. Instead, she had been banned from talking to a supposedly public agency and falsely branded a security threat to boot. The board had judged and convicted her without so much as a phone call or an inquiry.

On December 6, Ernst wrote back to Reid, demanding clarification. "First, are you alleging that by quoting what someone said to me in my email dated November 1st, 2005, I have somehow made a criminal threat?" Did the ban on contact apply to all issues, including explosive levels of methane? she asked. "Third, by what authority have you made the decision?" Fuming, the consultant posed more questions. How could a company conduct a credible noise study in response to a noise complaint with the noisiest facility turned off? How could a company report sound levels above what the board permitted yet still be deemed compliant? She also demanded that Reid tell Encana "to turn off its impacting machines until they are adjusted without violating my legal right to quiet enjoyment of my property."

Ernst's letter came back unopened with a bold stamp on the envelope. It read "Refused by addressee."

Hanneke Brooymans, a reporter from the *Edmonton Journal*, dropped by Ernst's house to talk about CBM development the day the lab results on Ernst's well water arrived in the mail. Ernst and the reporter opened them together. The resulting story made the front page of the *Journal*, along with a picture of Ernst and her dogs. The headline read "Tainted Water Lights Fire under Gas Fears." The story noted that Ernst had 44,800 parts per million of dissolved methane in her water and that the Canadian Association of Petroleum Producers considered 1 part per million a hazard in water passing through an enclosed space. (Alberta labs later changed reporting units to milligrams per liter.) The sample from the Lauridsens' farm contained 30,000 ppm of methane, while the bed and breakfast sample had 2,400 ppm. "I don't know what to do," Ernst was quoted as saying. "I'm still in denial." A spokesman for the EUB contacted by Brooymans said it was pretty rare for methane to migrate into a water well. A provincial spokesman blamed the finding on bacteria and said it was natural. "Ernst won't point fingers, but she knows there has been methane contamination of some wells from coal bed methane development in the U.S.," said the article. In fact, CBM development had sent methane on major migrations by then in both New Mexico and Colorado.

A day after the *Journal* ran the story, Encana's senior legal counsel, Carmelle Hunka, wrote to Ernst, stating, "It is Encana's observation that you are not interested in selling your property at this time." As a consequence, Hunka said, the company didn't have to meet Ernst's conditions for a buyout: full disclosure on fracking fluids used by the company, along with the names of other landowners around Rosebud dealing with methane contamination. But the lawyer assured Ernst that the company's drilling practices carefully isolated "groundwater from gas bearing sands or coals." Furthermore, the firm "investigates the ground water complaints we receive in conjunction with the well owner." As a courtesy, the letter added, the government usually got a copy of Encana's investigation.

By late December, she recalls, Jessica Ernst felt "weak at the knees." She routinely crawled into bed around 4 p.m. with Bandit and Magic, her laptop perched on a gaudy breakfast tray. It had become her new habit: as soon as the winter sun abandoned the day, she'd curl up with the dogs. Then, after briefly pretending that she lived a normal life, she would sit up in bed and attack her computer, looking for more scientific papers on gas migration and groundwater contamination.

Ernst saw herself and her dogs like Dorothy and Toto caught in a Kansas tornado. She had explosive levels of methane in her water well, but the regulator responsible wouldn't talk to her. Instead, based on something someone else had said, it had branded her a security threat. And to avoid Ernst's basic questions about fracking chemicals, Encana had closed the door on a buyout.

Ernst began to wonder if "the ugly piece-of-shit noise-maker" hadn't been a distraction all along. Maybe groundwater contamination was the real issue. What could the EUB and Encana be hiding? she asked herself.

It didn't take long before she found out.

SEVEN

Banished

IN EARLY JANUARY 2006, Jessica Ernst puzzled through a dense twenty-seven-page engineering study by an Edmonton-based company called Hydrogeological Consultants Ltd. (HCL). She read the report over and over again at her kitchen table, trying to parse its layered meanings. She had to decode a pipeline of jargon, including *turbidity, conductivity,* and *recompletion.* Water driller Chris Gerritsen had told her about the report after reading the *Edmonton Journal* story on Ernst's flaming water. "This might explain your problems," said Gerritsen, when Ernst dropped by his place to pick up a copy. "Encana had all kinds of trouble in the hills northwest of your place. They shoved everything down this well to try and stop the water but couldn't." Gerritsen angrily tapped the report. "Encana blamed me!"

The study, complete with illustrations and appendices, described the equivalent of a train wreck at a heavily fracked gas well. In February and March of 2004, Encana

had perforated and then cracked coal seams and other formations with nitrogen fluids on top of a hill above the home
of Sean Kenny, a local cattleman. Five months later, Kenny's
fifty-four-year-old water well filled with sand, bits of coal,
and foam. Thinking that the well's casing had failed, Kenny
hired Chris Gerristen to drill a new one that August. To Gerristen's dismay, the new well immediately coughed up more
black stuff, as well as suds as stiff as the head on a glass of beer.
Kenny suspected that "the activities of Encana" had caused the
groundwater contamination, and, at his insistence, the company hired HCL to do a study. During its investigation, HCL
looked at forty-six hydrocarbon wells that had been fracked
near Kenny's farm.

The investigating hydrogeologists zeroed in on Encana's
gas well 05-14, drilled just 1,200 meters northeast of Kenny's
place. The company had injected 70 million liters of frack fluids
at depths ranging from 125 to 400 meters into a series of thin
coal seams. "The shallowest perforated zone corresponds strati-
graphically [in the same rock strata] to one of the intervals"
in Kenny's two water wells, explained the report. As a result,
Encana's 05-14 had become a water geyser instead of a methane producer. Incredibly, the 05-14 gushed more than 8,000
liters of fresh water a day, enough to fill a hundred bathtubs.

The company squeezed a batch of cement in to block the
water. Schlumberger, the U.S. fracking king, describes squeeze
cementing, itself a form of fluid injection, as the act of pumping down cement to fill "a problematic void space" or fix a bad
well. When that failed, Encana made a second attempt. Excessive pressure during testing (a supervising error) cracked the ·
remedial cement, however, and in October 2004 the company
killed the well with a cement plug. That made it impossible

for the investigators to collect any further data on the well's contents.

To Ernst's horror, a detailed illustration accompanying the study clearly showed that Encana had punctured and then fracked into the aquifers supplying drinking water to Kenny's home and other wells in the Rosebud Valley. Ernst's own water well was only 3.3 kilometers (two miles) from the troubled well and in the direction of groundwater flow into the valley. She would later calculate, using the HCL report and records in the Alberta Groundwater Centre Database, that Encana had fracked 18 million liters of frack fluid over six zones directly into the aquifers. The report made no mention of the other cases of methane contamination that had come to Gerristen's and Ernst's attention around Redland. Ernst didn't know it at the time, but the HCL report presented a textbook case on what can go wrong with fracking. Norman J. Hyne, a professor of petroleum geology at the University of Tulsa, described the problem in the popular 2001 and 2012 editions of his *Nontechnical Guide to Petroleum Geology, Exploration, Drilling and Production*: "A well can be fraced several times during its lifetime and in some instances, however, hydraulic fracturing can harm a well by *fracing into water*. The hydraulically induced fractures extend vertically into a water reservoir that floods the well with water." The italics are Hyne's.

The phenomenon was already well known. Samuel Harrison, a geologist at Pennsylvania's Allegheny College, provided several case examples of the "hazards of contamination" from fracking in the journal *Groundwater* in 1983 and 1985. In northwestern Pennsylvania, Harrison reported, the "hydrofracturing" of tight sand had increased not only drilling activity but groundwater contamination. In one Rosebud-like incident, household water wells located near a valley floor bubbled with methane after a company drilled and fracked above them.

The fracked gas well in question extended down 4,000 feet, but the company installed its protecting steel pipe or casing to only 550 feet. The pressure inside the well forced dissolved methane into natural fractures that fed into the local aquifer. Several months later, methane blew the door off a landowner's pump house. Another resident could set water from his garden hose on fire. As Harrison documented, the gas driller hadn't noticed, near the gas well, a zone of rocks bearing lots of hairline cracks, but these natural fractures provided "avenues for rapid vertical migration of contaminants upward into the fresh water zone." The orientation of the fractures explained why some wells near the gas well remained unaffected while others 3,000 feet away filled with methane. Given these hazards, Harrison recommended that companies move their drilling sites "off valley wall onto valley floor or farther back onto upland if feasible."

Years would pass before Jessica Ernst would encounter Harrison's studies. In the meantime, she reread the HCL report with incredulity. Invectives poured out of her mouth as she grasped the report's full implications. At the community meeting in September 2004, Encana's Mark Taylor had categorically told two hundred people in Rosebud that Encana wasn't going to drill CBM or frack around the community. A month later, Taylor had assured the community that the company would never frack near a water supply. "They lied," says Ernst today. "They did what they said they wouldn't do, and then they lied."

Ernst knew the provincial rules on water diversions by rote because of her work in the oil patch. Under Alberta law, a company couldn't divert water from an aquifer without a permit, since the diversion might cause adverse effects to other water users, such as farmers or municipalities. When "a target coal zone is anticipated to contain and produce" fresh water, the law said, the company had to conduct "a Preliminary Ground- water Assessment containing baseline resource inventory data and other required information" before drilling. Ernst checked the records and found Encana hadn't done that. Moreover, the provincial guidelines for water diversion stated that if a company damaged anyone's water, they had to "resolve any allegation of impact" and fix it or supply an alternative source of water. Encana hadn't done that, either.

The report's conclusions, however, confounded Ernst. HCL hydrogeologists found that the contamination in Sean Kenny's wells, which included nitrogen, methane, and propane, was "not the result of Encana's activities." Instead, they blamed Chris Gerritsen for installing Kenny's second well using a poor design. (The conclusions so angered Gerritsen that he resigned from doing any water work for Encana.)

Nitrogen figured prominently in HCL's report. Encana had fracked the 05-14 well with nitrogen fluid, and lab tests found a nitrogen concentration of 17 percent in Kenny's well water. Despite that, the report's authors said they had no idea where the nitrogen in Kenny's water came from. They randomly picked a water well 185 miles (300 KM) away, in a place called Calmar, and found that the Calmar water well had similar levels of nitrogen. According to the report, that proved Kenny's nitrogen anomaly might be natural too. The use of such unrelated data seemed ridiculous, if not fraudulent, to Ernst. Why had HCL not compared nitrogen and methane data from local water wells? And why hadn't it done any fingerprinting of the methane and other gases? Ernst would later discover that Encana had tested and publicly reported a 30 percent concentration of nitrogen in its CBM well before abandoning it. HCL had had access to that data but hadn't included it in its report. As it turned out, HCL had found the same 30 percent composition of nitrogen in its first test of Kenny's well, but the company had thrown the data away, claiming it was a testing error.

Ernst delivered copies of the HCL report to colleagues in the oil patch for a second opinion. One went to Mike Watson, a genial, highly respected geologist who was an expert in migrating gases. Watson worked for the North American Oil Sands Company. The Canadian firm owned one of the largest pieces of real estate in the tar sands (1,110 square kilometers, or 275,000 acres, of land) and was making plans to steam bitumen out of the ground in the boreal forest. The company had earlier hired Ernst to do caribou protection planning as well as to help set up a three-year, multi-million-dollar stress hormone study, conducted by Sam Wasser of the University of Washington. The intent of the study was to predict how bitumen development in the forest might unsettle wildlife populations.

Watson had already told Ernst that the government's knee-jerk response to the elevated methane levels in her water was ridiculous. How could it blame bacteria and poorly maintained wells without doing the proper water tests or checking with Ernst first to find out if she shock-chlorinated her water well? Moreover, Watson said, the volume of methane in Ernst's water was too high and had changed too dramatically to be natural. He also told Ernst something interesting. Industry had discovered bacteria eating heavy oil in shales under Lloydminster, Alberta; after the bacteria started to produce methane, the gas leaked up wellbores and polluted aquifers, creating a costly problem. Watson knew all this because he had been part of the study group.

While perusing the HCL report, Watson extracted some additional facts from AccuMap, an industry database on every hydrocarbon well drilled in the province. Watson confirmed that in February 2004 Encana had punctured holes in the 05-14 well at 24 different intervals at depths of 125 to 419 meters from the rig platform and that "everything was fracked simultaneously in March 2 of that year." The company pumped cement down to seal the well off in July. When that failed, Encana corked it with a cement plug, supposedly solving the problem. "I detect the odour of a large bald-tailed rodent of the variety we're not supposed to have in Alberta," Watson wrote to Ernst. He and another heavy oil geologist agreed that the cement plug had probably forced liberated methane from the fracked zones into the path of least resistance: the Rosebud aquifer. In other words, Encana had made the problem worse.

When Ernst wanted to acquire more data about how shallow Encana's fracking had been around Rosebud, Watson showed her how to click on dots displaying a well site and then decipher the AccuMap data, which included basic drilling and fracking details. "It's deathly boring and will take forever," warned Watson. But over the next few years, Ernst searched more than 1,000 wells. She discovered that between 2001 and 2006, Encana had, in secret, fracked 190 gas wells above the so-called base of groundwater protection. She tallied the frack depths: 100.5 meters; 121.5 meters; 142.7 meters; 160.2 meters; and 164 meters. Ernst found fracks as shallow as 98 meters elsewhere in the province. Many of the frack jobs had taken place on top of coulees, in places where water wells were located in river bottoms.

All January, Ernst waited patiently for a phone call from Alberta Environment, sure that the explosive levels of methane in her water would prompt an investigation. The call never came. Instead, she got an increasing number of queries from concerned landowners. One was from Dale Zimmerman, a cattle rancher just north of Red Deer who had chatted with Ernst the year before.

Zimmerman's water had turned a silvery, cloudy gray in July 2005. A local water driller told the rancher that his water was so full of methane (75,800 ppm) that it could be set on fire. Two water wells on Zimmerman's farm, including one for his animals, also turned as fizzy as soda water after MGV (now Quicksilver Resources) punched in a CBM well almost a mile (1.6 kilometers) away. Zimmerman was forced to haul water four days a week, and he was greatly perturbed. After he complained to MGV, the U.S. firm hired HCL to do a study, which confirmed there was lots of gas in Zimmerman's wells but said it was all natural. Zimmerman didn't believe a word of it. He and his animals had drawn good clean water ("Gas Present: No") from a coal formation since 2000, and now industry had mucked that up. The rancher, who would lose thirteen cattle and one horse to polluted groundwater that year, told Ernst he didn't know what to do. The two talked about going public together and demanding a full and transparent investigation.

At the end of January, the EUB published a bureaucratic memo on shallow hydraulic fracturing operations. Directive 027 stated that the recent trend of fracking "shallow gas reservoirs" less than 200 meters deep using high fracture volumes

had damaged nearby oil wells. In addition, "information provided by industry to date shows that there may not always
be a complete understanding of fracture propagation at shallow depths." Nor did industry perform "rigorous engineering
design" for its fracks. As a consequence, the EUB banned companies from fracking at depths as shallow as 200 meters unless
they fully assessed the potential risks and shared those with
the board. The directive also promised random audits on shallow fracks. (Six months later, the board hadn't completed even
one audit.) In an effort to calm worried landowners, an EUB
press release asserted that "no fractures have impacted water
wells in Alberta."

139

The regulator had also tallied on one page some crazy "shallow fracturing incidents," though it never released the data
publicly. The document, which a landowner found in the
board's library, describes nearly a dozen cases of fracks traveling
far out of their intended "zone." In one event, a nitrogen frack
"made its way to surface exiting at a canal about 200 metres
away." After fracking one well at a depth of 287 meters, workers noticed nitrogen "exiting the ground" near an abandoned
well 20 meters away. Another company was fracturing a new
well about 20 meters from a sister well when "a pop was heard
and mist escaped" out of the sister well.

The EUB's botched frack job list echoed the findings of a
2005 article on fracking, written by a half-dozen engineers, that
appeared in Oilfield Review. The engineers once again admitted
that natural faults severely compromised the ability of industry to predict the geometry of their man-made fractures. The
cracks often shot out "of the intended zone reducing stimulation effectiveness, wasting horsepower, proppant and fluids
and potentially connecting up with other hydraulic fracturing

stages or unwanted water." Hydraulic fractures, warned the engineers, defied predictive models and displayed an "almost limitless range of complexities."

Ernst read the EUB compilation of bad frack jobs—subsequently tabled as part of her lawsuit—as a tardy admission of guilt and dashed off one of her trademark letters to the board. She found it sad, she said, that "only after oilfield wells were adversely affected by shallow fracturing did the EUB decide to set new requirements. Aren't water wells much more important to protect than oilfield wells?" Since the government had a public duty to protect groundwater, why didn't it conduct baseline groundwater testing first, so that contaminant trends could be tracked over space and time? asked Ernst. "Wasn't methane testing of area water wells vitally needed before fresh water aquifers were perforated and fractured into?"

The board didn't answer any mail from Ernst, and several urgent letters she sent to Neil McCrank, chairman of the EUB, seemingly disappeared into the ether. On what authority had the EUB banned all communication with Ernst at a time when her water well was explosive with methane? she demanded to know, to no response. Finally McCrank, a lawyer and former deputy minister of justice, tersely replied to Ernst's pleas: "I have asked Mr. Richard McKee of our Law Branch to deal with you directly with regard to your concerns and with your relationship with the Alberta Energy and Utilities Board in general." McCrank added that McKee would arrange a meeting to discuss matters more fully.

Ernst knew that McKee, who advised the EUB on policy, legal, and procedural matters, reported directly to McCrank. The lawyer had a reputation for toughness and for acting as the board's mechanic: a fixer of bad situations. When Ernst didn't hear from McKee after several weeks, though, she wrote McCrank again. In that letter, she asked why the chairman had ignored her questions about the ban: "Did you or the Board authorize this action and under what authority?" Ernst said in her letter she was sure "a man of your wisdom, experience and leadership" could answer her questions "in less than five minutes." McCrank never found the time.

After McKee's assistant left a message promising a meeting soon, Ernst wrote to thank the board for finally granting her an audience. In her letter, she asked why the board and McCrank had avoided her previous queries. She posed another question, too: "Did the EUB discriminate against me because I am a woman who dared to report the truth and express a neighbour's sad plea of despair at the EUB's apparent unwillingness to protect citizens and our environment from industry?"

In late February 2006, Rick McKee sent Ernst a blunt warning. He told the consultant there could be no progress on the substantive issues she raised "without re-establishing a climate of mutual respect." He said the EUB had been left with the impression, for better or worse, that "you are not interested in working with the Board in a constructive manner but are instead interested in using the Board process to further another agenda."

Ernst had no idea what agenda McKee was talking about. She fired back a letter saying she wasn't happy to be living in a home polluted with tainted water on an intensely impacted oil and gas field "where planning and cumulative effects are ignored." She had concerns, she wrote, for the health and safety of all citizens relying on groundwater "that is being risked by industry rushing madly in programs not always subject to rigorous engineering design" without "complete understanding of fracture propagation at shallow depths." She waited another three months for an answer.

David Swann, a Liberal member of Alberta's legislative assembly and the caucus's environment critic, had taken a keen interest in the growing CBM debate, and the lanky politician invited Ernst, Dale Zimmerman, and Fiona Lauridsen to present their water concerns at a legislative press conference on February 28 in Edmonton. Ernst, who hated crowds, reluctantly agreed. "I had no choice; the regulators just didn't do their due diligence," she later explained to a *Canadian Business* reporter.

The landowners emotionally shared their methane horror stories with reporters. Fiona Lauridsen, a Glaswegian with a tart tongue, told the crowd how her family had taken a shower before midnight mass and got their skin burned. The family could now set their water on fire, too. "The excuse that we need the gas is appalling. It's completely unacceptable. Our lives are being put at risk because the government wants to make money."

Dale Zimmerman and his wife, Brenda, a schoolteacher, explained that the gas had appeared in their water the day after MGV Energy drilled a nearby well. Their cattle wouldn't touch it. "You can literally see the gas coming out of the hose," Dale Zimmerman said. "It's a gray-white cloud." Health officials had told them not to drink or bath in the water. "Our five-year-old wakes up at 3 a.m. screaming about the water. Are the calves getting sick? Are her cats going to die?"

Ernst focused on the explosive levels of methane in her well water: "You keep thinking: 'It's water. It's safe.' I can't believe it's dangerous."

After the three had spoken, Swann asked Premier Ralph

Klein to halt CBM drilling until the government put in place stricter regulations. "It's totally unacceptable for the province to be allowing this CBM development with no groundwater management plan," said Swann. Ernst, she recalls, just wanted the existing regulations enforced.

The flaming water stories made headlines throughout rural Alberta, where most people sourced their water from wells. The furor caught the Alberta government off guard, and Premier Ralph Klein, a troubled alcoholic and a former journalist, promised he'd make industry do CBM right, because it was important to the economy. He also guaranteed Alberta citizens clean drinking water. "I am willing to extend that to the fullest extent," he told reporters. "Whatever is necessary to be done will be done." Alberta Environment Minister Guy Boutilier went even further and made a commitment to work with the families, including the Zimmermans: "I'll use every fibre in my body to assist this family relative to safe drinking water now and into the future... be it by natural flow or because of what is being asserted relative to what is taking place in the water supply." But as Ernst and other landowners would soon learn, the government had no intention of keeping its word.

The day after the press conference, a representative from Alberta Environment phoned Ernst about her water. Darren Bourget first asked Ernst to let Encana test her water, saying that was the normal practice. Ernst said no way. "I have documentation proving that Encana broke the law here," she told Bourget. "I don't want them anywhere near my land." For a minute, Bourget didn't know what to say. Then he offered to send a government water-testing team to sample for methane, hydrocarbons, and bacteria. Ernst agreed on the condition that the government send the results to Karlis Muehlenbachs,

whom Watson had told her was one of the best methane fingerprinters in western Canada. Bourget promised in writing to
comply with her request.

A few days later, Darren Bourget (to supervise) and a pair
of Alberta Environment investigators, Al Strauss and Leslie
Miller, showed up in Rosebud. A TV reporter, Kevin Green,
came along to witness their work. Bourget had promised Ernst
that the government representatives would bring her water
tanks and provide nonflammable water, but the investigators
arrived empty-handed and said no tanks were available, due to
industry demand.

Miller, a trainee, had never sampled water for hydrocarbons
or methane before. (Normally, representatives of the government watched as oil patch contractors did the work or didn't
show up at all and let industry do what it wanted.) As Ernst
observed the trainee at work, she offered some unsolicited
advice: "You are supposed to put the preservative into the vial
first, then fill with water and close the lid immediately to prevent the methane from escaping. Then turn it upside down.
If there is an air bubble, you have sampled incorrectly." But
Miller fumbled with the lids, Ernst recalls, and left the collection vials open for far too long.

Al Strauss, who checked out Ernst's well, did not wear
gloves, and his equipment appeared dirty. Ernst asked the
government investigators for their water contamination
investigation and sampling protocol before they did any more
testing. But she never got a clear list of accepted procedures.

As Miller took samples for hydrocarbons, bacteria, and dissolved methane and metals, Strauss checked out the static
water level and production in Ernst's water well in the chicken
barn. (Alberta Environment has never released the production

data to Ernst, despite a freedom of information request.) So much gas passed through the pump that it locked and shut off. "You've got a serious problem with your well," announced Strauss. "This is your land. You don't run cattle. That means you don't use enough water and caused this contamination." Ernst didn't say anything, but she wondered how a government investigator could reach a conclusion before he had gathered the evidence.

As an insurance measure, Ernst took her own water samples. The vials she filled—properly—came back from the lab showing three times more methane than the government-filled ones. It was now obvious that the government had no emergency plan for landowners and no water collection or safety protocol. Ernst had to call a plumber and arrange for the installation of fiberglass tanks on her own. It took the government a year to reimburse her.

While Alberta Environment bungled the water testing, the EUB, in its monthly newsletter, accused landowners of spreading myths about CBM. The article read like a thinly veiled attack on Jessica Ernst. The board insisted that CBM was nothing more than clean natural gas from an unconventional source. Although "the prospect of large scale development of a resource that isn't well understood" might generate suspicion, the article said, the CBM boom unfortunately provided an "audience for those willing to promote misinformation and mistrust." "Alberta Environment requires operators to obtain a groundwater diversion permit" when fresh water "is produced from any type of well including CBM," the article assured readers.

In the second week of March, Ernst left Rosebud to go on a three-day speaking tour organized by several landowner groups. Bandit and Magic came along and slept in her truck.

145

Billed as "Profit Zones or Sacrifice Zones," the hastily arranged tour promised to tell the truth about coalbed methane. Two American women who had flown up from the San Juan Basin also spoke. Gwen Lachelt, director of the Oil and Gas Accountability Project, talked about messy coalbed development in Colorado, comparing its impacts to hard rock mining. Tweeti Blancett, a well-known New Mexico rancher and Republican activist, explained how the industry had trashed her ranch with unending industrialization, toxic pollution, and a string of broken promises. "We have trouble keeping our few cows alive, because they get run over by trucks servicing wells or they get poisoned when they lap up the sweet antifreeze leaking out of unfenced compressor engines," she told Alberta landowners.

At a meeting in Trochu, Alberta, in the heart of CBM country, more than six hundred landowners turned up to hear the three women. Scores of people had to be turned away because there were not enough chairs. As cars converged on the hall that night, it looked like solid streams of light were approaching from every direction. Ernst started to sweat profusely and had trouble breathing. In a panic, she rushed to her truck and sat with Magic and Bandit. The prospect of being in a room with so many men filled her with a claustrophobic angst. Bandit leaned into her and calmed her down.

During her short talk, Ernst asked the farmers and cattlemen if they were landowners or guinea pigs. She didn't like being a guinea pig, she declared. CBM had turned her plumbing and water into a dire chemistry experiment, complete with whistling taps. "I have experienced sudden chemical burns to my skin and eyes. My water used to be soft and fabulous to bathe in and for washing. Now I don't get suds out of my soap and shampoo anymore."

She told the crowd about the HCL report and how Encana had fracked into the aquifer. She explained that methane can occur naturally in water wells, but usually at low levels that remain stable over time. She wryly dissected the fraudu- 147 lent noise studies, to sustained laughter, and the story of her banning by the EUB evoked gasps. "When we're reasonable, nothing changes," said Ernst. At the end she quoted an old Japanese saying: "Many fleas make big dog move." The crowd gave her a standing ovation. Members of industry and the EUB identified themselves by remaining in their chairs with crossed arms and scowls. As Blancett told Ernst, their body language said it all.

While Ernst was speaking to Albertans, Peter Watson, Alberta's deputy minister for the environment, left an urgent voice message on her business phone. Ernst had met with Watson a few days earlier, along with Environment Minister Guy Boutilier. Ernst had tried to show Boutilier the most damning sections of the HCL report, but the minister turned his back on her and refused to look at it. Members of Watson's staff had also behaved as though the report was radioactive, but a furious Ernst had insisted they at least write down the report's file number. During that same meeting, a government official had blamed Ernst, Lauridsen, and Zimmerman for causing their own problems by drawing too much water from their wells. It was an odd contrast to government investigator Al Strauss's conviction that Ernst's well suffered from too little use.

Now it appeared that one of the province's highest-paid civil servants wanted Jessica Ernst to stop talking about fracking altogether. In his message, Peter Watson—who is now the chair and CEO of Canada's National Energy Board—said that his staff had caught Ernst making critical comments about

Alberta's government at public meetings. "I'm very concerned about what you are saying," said Watson in loud, angry voice. He then hung up.

Oscar Steiner, a land advocate and one of the organizers of the landowners' tour, would later reflect that the Alberta government had done everything it could that spring to discredit Ernst. Spokesmen for the EUB even whispered to reporters in Nanton and elsewhere that Ernst had fabricated the water contamination because she was mentally unstable. "They had no idea how to handle her, none," recalled Steiner. "They had never dealt with someone as uncompromising as her. And they were in panic mode." One EUB official told Steiner he'd never seen anything as well organized as the landowner tour "except among terrorist groups." Was Ernst now a public enemy of the state?

EIGHT

Keys to the Bank

AT THE END of one of the Sacrifice Zone meetings, Shawn Campbell, a tall cowboy with a limp, approached Jessica Ernst. "We've got the same problem," declared the avuncular rancher. After Encana experimented with some horizontal wells under Campbell's property and fracked two shallow coal wells near his Ponoka ranch, the water had gone milky and now reeked of sour gas.

Gas testing found H$_2$S, methane, ethane, propane, butane, and pentane in Campbell's well. The government reluctantly investigated the incident and informed the Campbells that they had "the only affected water in Alberta." Shawn and his wife, Ronalie, devout Christians, would form a "deep thermogenic relationship" with Ernst in the years to come, as their cases dragged on.

The Campbells had dealt with industry cordially for years and had nearly thirty oil and gas wells on their property. But as soon as lab tests identified deep gas from energy wells in their water, everything changed. The Campbells would test one company's "open door" policy by visiting the company's Calgary office to tell their story. The president listened to the Campbells' concerns about gas migration and fracking for only a short while. "And then," Shawn recalls, "all of a sudden he got up and he drove his fist into the table and he said that's enough of this and we won't talk about it anymore." Following his outburst, the president walked out of the room.

Industry and government had adopted the same mantra, it seemed: methane in groundwater was as natural as sunshine. To support the claim, Alberta Environment released data showing that 906 water wells in the province contained gas "assumed to be methane" and nearly 26,000 water wells had been drilled into coals. The detected methane had seeped from the coals, reasoned the government, and it had been doing so for decades. As one industry lobbyist later put it, "If it [methane] is present, the presumption is that it's naturally occurring."

Jessica Ernst didn't believe the government's selective evidence, and she continued doing her own digging. On the EUB's website, she found an incriminating presentation by Jim Reid, the senior EUB official who had banned Ernst from contact, that described shallow fracks as high risk, because gas drilling was occurring closer to the surface and closer to water wells. (The presentation has since been removed from the regulator's website.) Ernst also found a presentation by Heather von Hauff, an Alberta Environment hydrogeologist, in which von Hauff disclosed that Alberta had 500,000 water wells. If 906 of those wells were contaminated with methane, as Encana and the government had cited, that was a figure of 0.18 percent, Ernst realized. That made gas in water wells a rarity, not a common, natural occurrence.

Next, Ernst hired a hydrogeologist to review 2,300 historical water-well records within 50 square kilometers (about
12,000 acres) around Rosebud. For decades, the government
had maintained a public registry on the province's water wells. 151
The record for each water well had a line that read "Gas Present"
with an option to write either Yes or No. Prior to the shallow
fracking boom, only 4 of the 2,300 wells—0.17 percent—had
written Yes to a gas that might be methane. That directly
matched the findings, Ernst discovered, of a little-known 1995
study by Alberta Environment for the Canadian Association
of Petroleum Producers. For that study, Alberta Environment
reviewed 24,000 historical water-well records prior to oil and
gas development in the Lloydminister area. The study found
that only 17 wells reported gas present prior to the industry's
arrival. After industry started to puncture and frack the landscape between 1960 and 1995, the number of contaminated
wells rose to 41. That meant that, by 1995, 0.17 percent of
24,000 water wells in the heavy oil area northeast of Rosebud
contained small amounts of methane—usually less than 0.05
milligrams per liter. (One monitoring well near a leaking oil
well, contained 19.1 mg/l.) The small percentages found in the
hydrogeologist's review retained by Ernst and the study done
by Alberta Environment for CAPP told Ernst that government
and industry claims about common methane contamination
were false.

The two studies confirmed for Ernst that government
claims about natural methane were vile nonsense. Water
wells with gas in them appeared infrequently, and those with
the highest concentration of methane were usually found near
leaking energy wells. A 2003 regional groundwater assessment for Wheatland County that included Rosebud hadn't

mentioned methane at all. A groundwater study for a neighboring county "did not report dangerous or explosive levels of
methane as a naturally occurring phenomena." In fact, none
of the government studies Ernst had found by then reported
methane levels as high as the 30 or 66 milligrams per liter that
had been measured in Rosebud after the shallow fracking.

As concern about unconventional drilling heightened
among landowners, the Alberta government mandated
baseline water testing for CBM in April 2006 but only
for fracs in the freshwater zones. All water wells within a
six-hundred-meter radius of such CBM wells now had to be
tested for the presence of methane. However, testing for fracking fluid additivess such as BTEX or heavy metals—all critical
indicators of contamination—wasn't deemed necessary. To
Ernst, the new directive was too little too late. Industry had
already conducted its experiments and dirtied the waters.

Senior staff at Alberta Environment held several private
emergency meetings about CBM in May as a result of all the
media about flaming water wells. Snippets from the meeting
notes, obtained by Ernst in a lengthy and bitter freedom of
information request process, reveal a state of confusion, dis-
order, and misinformation. "People's expectations of Alberta
Environment" were "too high," the notes said, because "they
won't do anything." The EUB had no plans to do its own testing, and it didn't trust isotopic fingerprinting, either. One staff
member divulged that "Encana has also been drilling people
new [water] wells & giving them $."

In the notes, Ernst's case was described as the most vexing
one. She wouldn't let Encana do the investigation ("the normal
procedure"), staff members said, and now she wanted the government to do a full cumulative impact assessment and "won't

help even to collect data." Another note read, "Still feel bad for Jessica because she clearly has bad water."

That same month, Ernst had a bizarre encounter with government. On a hot day in early May, she drove over to Debbie Signer's bed and breakfast in Redland, situated in the river valley just a few miles west of Rosebud. Signer, a single mom, had started her Field of Dreams B&B in 2002, complete with a brand-new water well. Like many of her neighbors, Signer had a cistern. Wells in Redland delivered water so slowly that residents often installed large concrete cisterns to make sure they had enough water to meet demand at any point of the day.

Signer's new well started to have problems after Encana drilled and fracked several shallow gas wells nearby. Strange pink growths also appeared in her toilet. Later, her water began to fizz and turn milky white. The EUB didn't respond to Signer's complaints for more than six months. When Alberta Environment finally agreed to investigate, Ernst and Signer agreed that having some witnesses present might keep the operation honest. Fiona Lauridsen, one of Signer's neighbors, showed up too.

While waiting for government investigators, Ernst, Lauridsen, and Signer chatted in the kitchen over coffee with a Calgary reporter named Jeremy Klaszus. The women swapped tales about the industrialization of the landscape and contaminated water. Methane seemed to do a fine job of staying in the coal before it was fracked, Ernst pointed out. "Fracking is, after all, supposed to release methane. If they don't frack a CBM well, they don't get gas out. So why would a water well get gas out if we don't frack, and we never had it before? And then they frack our aquifers." Klaszus, who wrote for *Alberta Views*, taped the conversation.

When Alberta Environment arrived, Signer served the investigators, Leslie Miller and Kevin Pilger, a coffee. Pilger, a big man with a heavy mustache, had worked nearly twenty years for the government investigating environmental law-breakers. He had begun his career with Alberta Environment as a field technologist, specializing in air emissions and ambient air monitoring, and now he was doing water studies. The group sat tensely around the table.

Ernst, as direct as ever, got down to basics. "When is the government going to hire more manpower to do proper baseline testing in the area?" she asked. Pilger agreed that was a good question. "Maybe it would make sense to slow down development so government could catch up?" continued Ernst.

Pilger did most of the talking and explained how things worked. Industry controlled the pace of development, he explained, not government. And the government encouraged industry to mostly police itself. The government didn't want to be Big Brother in these sort of investigations, said Pilger. "We want people to accept responsibility for their actions."

"Including Encana?" asked Ernst.

Yes, Pilger said, including Encana.

"Our first call after a complaint is to the company," Pilger added. "We say, 'Okay, I've got a complaint. What are you going to do about it? We don't have the resources to be addressing all of these complaints. If you caused the problem, you find the answer.'"

The three women were stunned. They couldn't believe their government had such a spineless approach to enforcement.

"If you've captured a bank robber in the act, does the RCMP then give him the keys to the bank and say, 'Here you go, John. Go fix the problem. Put the money back'?" asked Ernst.

"That is not a fair analogy," stammered Pilger.

Pilger then emphasized that he could only accept what companies like Encana told him, and they had promised not to frack aquifers. Signer walked to her kitchen sink and filled a glass of water. The tap hissed, and the glass gave off a white mist. She placed it in front of Pilger. "You look at that and you can see vapor coming off it," she said, stating the obvious.

Pilger explained the government's position. "It's been well known in certain parts of Alberta for years that there is natural gas in groundwater. There is gas, and different kinds, everything from methane to carbon dioxide, in groundwater."

"And where's the baseline data?" asked Ernst.

Most of it was anecdotal, admitted Pilger.

"So you are allowing something like CBM to go forward without getting the real baseline data," Ernst continued.

Lauridsen jumped in. "Why can Encana and the government use anecdotal evidence but I can't?" she asked. "I did not have this problem in my water before, and now I do. That's anecdotal."

Ernst told Pilger that, given the absence of baseline data, landowners felt like guinea pigs with two legs.

Pilger said he had "no response" to that kind of comment. Miller just sat and listened.

After finishing his sampling at Signer's place, without using gloves, Pilger drove to Lauridsen's property, just a couple of miles away. The women followed. There Ernst asked Pilger (Miller had gone elsewhere) if he had looked at the public data on the perforating and fracking done around the Rosebud area.

Pilger said he hadn't; he trusted Encana when it promised it would only frack deeper than 600 meters below the ground surface, or way below the level that groundwater flowed. Ernst asked Pilger if he had looked at Encana's own frack depth data, which showed the company cracking coal and sands at 125 meters. When pressed on the matter, Pilger accused Ernst of breaking into the EUB's offices and altering Encana's perforation and fracturing data.

"We can't look at Encana's data at the EUB, because you broke in and altered it all," he told Ernst.

"You mean I changed thousands of documents by myself?" said Ernst, incredulous.

Nothing if not dogged, Ernst then asked Pilger if he had read the HCL report, which neatly illustrated how Encana had fractured into groundwater. Pilger said he couldn't read the HCL report because Ernst had fabricated that study, too. She calmly pointed out that the report had been professionally sealed and had two signatures on it. Ernst probably forged those signatures and the seals too, Pilger said. At that point, Lauridsen recalls, she went inside her house thinking she might faint. She later characterized the episode as totally surreal and described the regulator as a lunatic: "I've never seen a guy in such a state of denial."

Years later, Ernst would relive the unsettling encounter for audiences in Canada, Ireland, and England who were concerned about global frackers invading their backyards. She'd show engineering seals on the HCL report and then a slide of Wonder Woman, sarcastically explaining that she'd morphed from an ordinary scientist into a superhero capable of altering masses of regulatory data. It must have been Wonder Woman, too, who had forged the HCL report and the engineering seals. "One of the positive things about getting fracked is that you can become Wonder Woman," Ernst would joke. The story invariably produced a wave of uncomfortable laughter.

Ernst encountered Kevin Pilger again at Debbie Signer's place in June. The May tests from Signer's cistern had shown no detectable bacterial counts but lots of methane (27 mg/l). Now the government was back to do more sampling. Still 157 upset by the tenor of the May visit and the government's reluctance to deliver safe water, Signer had asked Ernst to come and witness again. This time, Ernst brought along seven First Nations elders from the Yukon who were on a tour collecting facts about how fracking affected the land and communities. Ernst brought along Bandit and Magic, too.

With growing unease, Ernst stood about fifteen feet from the well to watch Pilger do his tests. Pilger used a rope instead of a sterile chain, along with a sampling canister. It was a hot and muggy day. Pilger rested both the rope and the canister on the muddy ground, where Ernst had noticed a generous amount of gopher shit. She respectfully pointed out that disturbing fact to everyone watching, announcing that the government was likely not only collecting contaminated samples but introducing bacteria into Signer's well system. Pilger said nothing but continued working.

Later that afternoon, Tadzio Richards, a visiting documentary filmmaker, noticed Pilger poking around Ernst's yard. Did Ernst know a government inspector was on her property? Richards asked her. Ernst rushed outside, wondering why the compliance officer hadn't come to the front door to announce his presence. When she asked what Pilger was doing, the officer said he was taking GPS data to determine the location and elevation of her well; he planned to compare that to the fracking data from Encana, he said. (The government later refused to release the collected GPS data to Ernst.) Pilger also admitted that Ernst had been right about the dirty sampling at Signer's place. He had thrown those samples out and taken new ones, he informed her. Ernst knew that wouldn't be possible without shock-chlorinating the whole system, but she kept her thoughts to herself.

One night shortly afterward, Ernst had a premonition and woke in a cold sweat. She called Debbie Signer and told her she thought the government was trying to set Signer up for being vocal about her water contamination. Through Pilger's sampling, the government had introduced bacteria into Signer's well. That meant there was an immediate public health hazard for Signer and her guests. But Ernst worried also that the government might follow up Pilger's visit by sending over a health inspector to test Signer's well. If they found E. coli, they would shut down Field of Dreams and, with it, Signer's methane well complaint.

Ernst offered to come over and shock-chlorinate Signer's well, and Signer immediately agreed. It took a few hours, but the women ran chlorinated water through every tap, toilet, and water outlet. At ten to five that day, a government health inspector arrived at Signer's place and said he wanted to sample her water. (The collection of bacterial samples on a Friday is regarded as bad practice for good reason: the samples sit untested over the weekend, growing more bacteria.) Signer cheerfully told him that her well and cistern had just been chlorinated. Signer never got back any test results, and she suspects the inspector threw the sample out his car window on the way home.

Signer did receive a frosty letter in August from Alberta Environment. "The bacteriological impacts to your water well and cistern are related to water well maintenance and not to coalbed methane drilling activity," the letter stated. Water delivery, which had started in late June, would now be suspended. Flabbergasted, Signer replied in writing that she had seen no testing results whatsoever. Moreover, she noted, Alberta Environment had "done sampling with Mr Pilger repeatedly dropping a muddy rope down my well which lo and behold collected E-coli?" Signer added that the "Alberta government is suppressing the chickens in the henhouse while the foxes run free." She was more right than she knew; later testing would find fifty chemicals in Signer's water, along with an incredible volume of methane: 110 mg/l. (Industry warned that there was a risk of explosion with just 1 mg per liter if the water passed through a confined space.)

159

Just weeks before Kevin Pilger's first visit to Signer's B&B back in May, another landowner, Bruce Jack, had called Jessica Ernst to ask for help. Ever since industry had drilled and fracked a dozen oil wells within about half a mile of his farm in Happy Valley, Jack had had problems with methane and ethane in his well. So much gas was pouring up the well that it had punched out the walls of the pump house, Jack told Ernst, and one day an explosion had scorched the inside of the shack. The oil company had reluctantly hired a water supply company to professionally install a venting system, yet a spurt of methane-frothing water still pissed out of a pipe. Neither regulator wanted to get involved, explained Jack. The EUB said it was Alberta Environment's responsibility, and Alberta Environment bounced everything back to the EUB. When Jack objected to more drilling near his farm, because his water was already fouled, the EUB told him that "contaminated groundwater was not a ground for objection."

Fed up, Jack wanted to go public. "Will you help me?" Ernst said yes. She drafted a news release, but it never got sent. Days later, two gas-in-water experts hired by industry showed up to test Jack's water. Just as Jack entered the pump house, an explosion blew it to pieces. The next thing Jack remembered was standing seventy-five feet from the building, tearing flaming clothes off his body. All three men ended up in hospital. Jack, who suffered third-degree burns to his face, back, and arms, spent a month there. Peter Watson, deputy minister of environment, promised Jack water delivery and an investigation if he agreed to remain silent. Jack's family reluctantly agreed. To this day, no regulatory record of the event exists in the public domain.

Not all of Ernst's neighbors welcomed her scientific grit. By early summer her persistent questions about groundwater contamination had unnerved many residents around Rosebud. Indebted farmers needed the income from leasing their land to CBM drillers, and they thought Ernst's speaking out might jeopardize that revenue source. Some told her right to her face to shut the fuck up.

Senior managers of the Rosebud Theatre also accused Ernst of "misinformation." Encana had promised $150,000 to the theatre in 2004—as Ernst put it—"to not only divide and conquer but to buy people's silence." The theater, however, regarded the promised funds as a godsend. One day Bob Davis, the director of the theater, phoned Ernst. He wanted her to tell the press that she had been wrong about the aquifers; patrons were now bringing their own water to the dinner theater. Davis was captain of the theater, he told Ernst, "and you are making me steer this ship into the shore." Ernst listened to the tirade, then told Davis, "I'm not lying for you or anybody. You could do the right thing and say to patrons that it is a good thing that [they] brought bottled water." Davis didn't see it that way and later recalled it was difficult to know who to believe at the time. By the end of the year, many people around Rosebud had started to shun Ernst "as that crazy woman." Some called her an extremist. Encana even asked the local paper to stop printing Ernst's letters.

Rattled by the huge crowds the Profits Zones or Sacrifice Zones tour had attracted, and by the growing number of water-well complaints, the Alberta government decided to hold its own "information sessions." According to Environment Minister Guy Boutilier, the meetings would "ensure landowners have access to all the facts about how their groundwater is protected during coalbed methane development." The government took its road show to twelve rural communities, nearly everywhere Jessica Ernst had given a talk. At every meeting, members of the EUB, Alberta Environment, and the Canadian Society for Unconventional Resources told landowners not to worry: groundwater contamination just didn't happen in the province. The speakers claimed that "industry meets or exceeds requirements" and warned citizens not to believe everything they heard in the media. More than seven thousand CBM wells had now been drilled, they said, with "no instances where CBM has caused contamination to a water well." Some claimed industry had *never* contaminated groundwater.

The government-sponsored meetings encountered skeptical and often hostile crowds. At one session in Strathmore, a government official confessed, under questioning, that even though industry had drilled 300,000 oil and gas wells and a third of the province's population depended on groundwater for their drinking supplies, the province had no standard procedure or protocol for investigating cases of groundwater contamination. The company responsible for the damage

performed a study only at the behest of the government, and
the whole process might take a year. The revelation astounded
the crowd. A veterinarian called the government approach
unprofessional, and a water-well driller compared the con-
flicted arrangement to a coroner murdering his wife and
then being allowed to conduct his own investigation under
police supervision. At a meeting in Torrington, Alberta, an
Encana official defended the system, saying it worked perfectly. Encana had checked out thirty-six water-contamination
complaints to date, he told the crowd, and had found it wasn't
responsible for any of them.

What had already happened in Colorado, Wyoming, and
Alabama could never happen in Alberta, the touring officials
insisted. The Americans were two decades behind world-class
Canadian regulators. If that was the case, asked one brazen
landowner, why was the EUB waving around the discredited 2004 EPA fracking report as though it was good science?
Where was Alberta's own fracking study?

Some government and industry statements made on the
tour stuck out like scarecrows in a garden. An industry lobbyist
declared that shallow fracks "almost always stayed within the
coals," while a regulator claimed that "water does not migrate
underground." Officials with Alberta Environment repeatedly
blamed methane fouling on a sudden epidemic of bacterial
growth caused by sloppy water-well maintenance, saying link-
ages to CBM drilling or industry didn't exist. One bureaucrat
declared there had been gas in Alberta water wells "since the
start of time."

Lots of farmers asked why the government hadn't done
proper baseline testing, since that was recommended by the
Canadian Council of Ministers of the Environment. In 2002,

a groundwater paper prepared for the ministers had concluded that unconventional natural gas drilling posed such
a threat to groundwater quality and quantity that the nation
needed "baseline hydrogeological investigations in coalbed   163
methane...to be able to recognize and track groundwater contaminants." None of the officials had an answer to that. Jessica
Ernst attended most of the sessions and took notes. "Why has
the regulator allowed the oil patch to risk Alberta's groundwater when safeguards could so easily have been implemented
before the experiments took place?" she asked at one meeting.
She didn't get an answer either.

Shortly after the industry tour dismissed groundwater
complaints as mistaken, David Pryce, vice president of the
Canadian Association of Petroleum Producers, told the Calgary Herald there were no "demonstrated problems" with CBM;
if there were, the industry would fix them. "This is just another
form of natural gas," he said, "and we are well-experienced
at developing and producing natural gas." Alberta Environment Minister Guy Boutilier said no CBM moratorium was
necessary: it was all about maintaining the delicate balance
between economic progress and environmental protection.
Roger Clissold, the president of hydrogeological consultants
ltd., explained the real problem was not carpet bombing of the
landscape with shallow fracked wells, but unregistered water
wells: "In my opinion the biggest threat to the groundwater resource is an unregulated water well industry and not a
highly regulated oil and gas industry."

David Schindler, a world-famous water ecologist at the
University of Alberta, saw things differently: "It's the same old
story for this province," he told the Calgary Herald. "Balls-to-
the-wall development without any background homework at

all... Our knowledge of groundwater resources in this province is disgusting. Without any detailed knowledge, I don't see how we can intelligently design any sort of coal bed methane development."

The rural meetings didn't ease the pressure on the Alberta government, however, and for good reason: the authorities withheld some important information. No official divulged, for example, that the province had hosted an entire conference devoted to CBM in 1989. At that gathering, researchers had warned that "out of zone fracture initiation merits further investigation." Nor did touring officials disclose that the EUB had set up a task force with the goal of monitoring practices and setting policy for CBM drilling but had disbanded the group in 1995. Harold Keushnig, a retired manager, would later describe the EUB's failings to the *Calgary Herald*: "I guess if we would have really been thinking ahead, we should have realized the coals in Alberta are shallow and those are the ones that would be developed first—and that there would then be an effect on shallow groundwater supply... We, the industry, maybe should have thought about that at an earlier stage."

Jessica Ernst finally had her showdown with the EUB in early June. The extraordinary encounter, most of which Ernst taped, unfolded like some absurd Monty Python sketch. The meeting took place after a four-month flurry of emails between Ernst and Rick McKee, the board's legal adviser. McKee repeatedly warned Ernst that none of her noise or groundwater concerns could be addressed by the board until her relationship with the regulator had been properly ironed out. The "first priority is the establishment of a framework for constructive engagement... without that little else can be accomplished," said one email.

In return, Ernst demanded that McKee actually answer a question. For example, why had the regulator allowed "energy industry approvals processing to escalate at the expense of human rights, public health and safety, and the environment?"

The lawyer and the landowner held their parley in the downtown Calgary office of Liberal MLA and environment critic David Swann. Swann, a former doctor, had continued to take a keen interest in Ernst's case. A gentle optimist, he opened the meeting by suggesting that only "more responsible investigation and more open dialogue" could pave a way forward. Ernst handed McKee a Tibetan flag.

"What's this," asked a suspicious McKee.

"A peace offering," Ernst said.

The meeting, which lasted nearly two hours, proved to be both a failed interrogation and a sharp dialogue on the nature of justice and power. It also suggested how tenuous the relationship between energy regulators and landowners would become as the industry fracked more landscapes and put more water sources at risk.

Ernst recorded half of the conversation before she ran out of tape; she hadn't expected the meeting to last as long as it did. Although she had requested that two local journalists, Jeremy Klaszus and Tadzio Richards, be present, McKee had told Klaszus not to show up. McKee wouldn't allow Richards into the room, either, even though the lawyer had initially agreed to witnesses. As a consequence, David Swann mediated the affair. Much of the heated discussion concerned noisy compressors, "the Wiebo Way," and what McKee repeatedly described as Ernst's public shaming of the board.

McKee, a burly fellow with unkempt hair, got right to the point. "Late last fall the board took a decision internally to discontinue further discussions with Jessica."

Ernst interrupted. She wanted to know if the whole board, a gaggle of white, middle-aged male lawyers and engineers, had made the decision to ban her, or if it was the whim of one civil servant. McKee replied that when a letter left the board, it was a board decision. "Whether the board actually sat and scratched its collective chin about it is another thing. I mean, the organization speaks with one voice... If a corporation writes you a letter, you could take that to mean that the corporation stands behind that," he said.

"What's the goal of this meeting?" asked Swann.

McKee said it was important for Ernst "to understand where we're coming from, what happened, why it happened." He was there to state the organization's concerns about her behavior and to try to sort things out: "I don't want to continue the current situation with Jessica's voice in the wilderness." McKee said he also wanted to know "what her goals are, what is it she wants and what role she wants to play," so that communication could go back to normal.

Changing Ernst's status with the board would "require a couple of things," noted McKee. "Number one is going to require recognition of the fact that certain behaviors, for want of a better word, OK, are counterproductive to playing a role in the overall board process." The board, emphasized McKee, was concerned about its protecting itself from criminal threats and ulterior motives.

The lawyer then peppered Ernst with questions about who she really represented and why she had raised such a stink about the compressors. It was one thing for her to involve herself in board business, McKee said, but another to use that "as leverage for a broader agenda. You can't expect the board to cooperate in its own barbeque."

Ernst replied that she had lost her trust in the EUB long before the "banishment" letter. She explained that she represented not only herself, but the community of Rosebud, where compressors had been installed without proper noise muffling. She also acted as the informal spokesperson for a small group of landowners in the badlands around Drumheller known as the Valley Group, she said. She explained that she didn't take any payment from landowners and didn't represent individuals. "Almost everyone who phones me asking for help, I tell them that they're much more powerful in resolving the problem if they learn how to do it themselves rather than if someone does it for them. My grandfather taught me to do it yourself."

"Just leave that aside," interrupted McKee. "I just want to go right to the nub of what triggered this banishment as you call it, OK?"

"What is it?" asked Ernst. "It's a threat. It's an intimidation. You cc'ed the letter to the RCMP and the Attorney General."

The board "felt threatened by you," explained McKee. "Legitimately felt threatened by you. When I say legitimately, they actually felt threatened by you, whether the threat was real or not. They took it as real."

Ernst calmly replied that she felt "incredibly threatened by the EUB and Encana."

"There were comments attributed to people within your circle... that talked about resorting to Wiebo-like tactics," retorted McKee.

The Wiebo Way had nothing to do with violence, Ernst responded. "How is a regulator in a democratic society allowed to make a judgement like this, without a trial or a hearing to ask 'What is the definition of the Wiebo Way?' "

167

"I'm asking," said McKee.

"But shouldn't you ask such questions before you banish people as a security threat?" repeated Ernst.

McKee explained that the board had "bent over backwards" to solve the compressor-noise problem and that there were "extraordinary measures taken to try and get at least to the point where we could agree where we disagree. That never happened. This short-circuited the process, but when a government agency or quasi-government agency…feels that there are threats implied or direct to their personal safety, there's no right that you can use to trump the fact people have the right to their personal safety."

"Don't you think, though, it's very important to find out whether there was an implied threat or not?" asked Ernst.

Ernst added that she had quite a sense of humor, and that the board's leap of judgment about the Wiebo Way was not only foolhardy but stupid. "Whatever happened to everything that Wiebo has done on alternative energy? The Wiebo way represents that we can't do these fossil fuels anymore. It's driving us crazy. It's destroying the groundwater. Destroying the land. People can't sleep. We can't continue to do the resource extraction the way we are doing it."

At that point, David Swann suggested that maybe the "Wiebo Way" didn't mean industrial sabotage at all.

"The Wiebo Way has nothing to do with Wiebo," Ernst added. "The Wiebo Way is about using French-fry grease as an alternative fuel."

McKee expressed astonishment. "Are you suggesting to me in all seriousness that the average Albertan hearing the phrase 'Wiebo way' is going to think of French-fry oil in the car?"

Actually, no, replied Ernst. "You know, what they think of first is wind energy. 'Cause the first thing he did was wind."

(After their war with industry, Ludwig's family installed a windmill and made diesel fuel from cooking grease.)

"When you use a phrase like that…"

"I didn't," snapped Ernst. "One of the members of the group did." 169

"Fair enough," replied McKee.

Ernst added that a member of the Valley Group had made the comment about Wiebo because the relentless CBM compressor noise was driving him crazy.

"And you think he was talking about French-fry grease?" asked McKee.

"He was talking about stopping violent resource extraction that unsettled people's lives," shot back Ernst.

McKee suggested there was only one interpretation for the Wiebo reference, and it had to be blowing up oil and gas facilities.

"No one thinks violence is a good way," replied Ernst. The more than three hundred people on her email frack news and the CBM watch lists, she said, know "that CBM's really a dirty extraction. It comes with a lot of serious impacts to the people living on the land." Ernst repeated that the board should have consulted with her about the meaning of the email before banishing her.

"Fair enough," said McKee.

The verbal jousting continued like some highly competitive tennis match.

McKee defended the board. "The intent to intimidate you, what it was trying to express to you in the strongest possible terms, that there are certain lines over which, if you wish to go, off you go. But you'll do it without the board. The board will not continue in any sort of engagement with someone who is advocating violence."

"I'm not advocating violence."

"Would we sit here talking if I believed it?" The lawyer then asked Ernst if she would stand before the board, a semijudicial body, and duly swear that "the advocating of violence was the furthest thing from your mind" and that she could never advocate violence.

"Absolutely," Ernst popped back. She didn't even own a gun to kill gophers, she said, and yet she lived smack dab in the middle of gopher country.

McKee then moved onto another pressing issue: the sharing of board information with the public. Prior to banning contact with her, the board had sent Ernst a draft copy of revisions to the new noise-control directive. The revisions, McKee said, had proposed lowering standards so that CBM companies would have an easier job of placing their compressor stations.

The proposed revisions also had the effect of legalizing the false noise studies by granting industry a 5 decibel noise increase, Ernst said—which just happened to match Encana's noncompliant noise levels at Rosebud. Ernst thought the proposed changes in the industry's favor were an affront to good public policy, she said, and so she had eviscerated them in letters to the regulator and on her mailout lists.

McKee didn't consider such behavior "constructive engagement," he replied. "The next thing we see is that it is being spread on the Internet as if it were board policy and the board is getting slagged from all sides because of that... If you've got criticisms, bring it back to us... This is where I need to be absolutely frank."

McKee added that if Ernst had plans to effect some sort of sea change on the political system or the legislative direction of the province or the fossil fuel industry, that was fine. "That's your right, but at the same time you can't expect the board to willingly stick out its neck and allow you to lop it off every chance that you get as a public example of your broader concerns."

The information on the directive needed to be shared with the public, Ernst said.

"But do you understand how annoying and how frustrating that could be to the board," insisted McKee.

"What about ordinary citizens and their fate in the wake of mass drilling?" retorted Ernst.

McKee next accused Ernst of belittling the board. "I don't want to make it sound as if its people are a bunch of sensitive... but at the end of the day you seem to be attempting to humiliate the organization. I mean, if that is your intention, good on ya. But don't expect us to help you."

Ernst replied that she "had no intention of humiliating anybody," repeating, "I've lived now for almost two years with this noise."

David Swann quietly interjected, "We're talking about a progressive breakdown of trust." What had happened between Ernst and the board, he said, was a microcosm of what was happening between the board and landowners across the province.

If Ernst felt fundamentally that the board was not capable or interested in running a fair process, demanded an increasingly frustrated McKee, why was she continuing to engage with it?

"I don't know if the board is doing that on purpose," said Ernst. "When you're on a Ferris wheel, you don't really realize you're moving. You're not seeing the whole picture. You're seeing your world and industry's as much more powerful."

"You are too intelligent and too capable to bash us," replied McKee.

"I've learned that being reasonable doesn't work," retorted Ernst.

After her tape ran out, Ernst remembers, McKee bluntly asked her what it would take to get her to leave the province. She leaned forward and glared. "As soon as you start doing your jobs, I'll happily leave."

As the conversation continued, Ernst slammed the board for its poor cumulative-effects planning and its nonexistent emergency response to groundwater pollution. McKee claimed that hundreds of people had emailed the board saying Rosebud's water was always fouled with methane. "Show me the data," said Ernst. Yet McKee never produced any data on the matter.

In the end, McKee promised to get Ernst an audience with the board and to "reinvigorate her into the process."

Following the meeting, Ernst concluded that McKee's orders had been to shut her down by getting to her to admit to having bombs and making criminal threats. "The abusers tried to break me," she recalls, "but it didn't work. The only bomb I had was the bomb in my water."

Ernst never got an audience with the board. Nor did McKee pursue her offer to share in the costs of hiring an out-of-province mediator to patch things up. The broken promises convinced Ernst that the meeting had had nothing to do with restoring her standing with the board: "It was just about trying to scare me into allowing myself to be controlled."

That same month, Gerry Protti, the rotund vice president of Encana, characterized Ernst's water claims as baseless. He told a BC-based magazine investigating the company's million-dollar donation to the University of Victoria that "we can demonstrate that we've never impacted an aquifer in our drilling." Asked about Ernst's flaming water, Protti replied, "That's actually fairly common." Methane occurred naturally, and, furthermore, the water was safe to drink. 173

A year later, Rick McKee departed the EUB as a major scandal enveloped the regulator. Leading up to a public hearing on a controversial transmission line, the board's security branch—the same men who had overseen Ernst's banning—hired four private investigators to mingle with landowners and provide "a covert security presence" as well as "intelligence gathering." The spies, who pretended to be upset landowners, monitored conference calls, requested documents, and eavesdropped on conversations, including privileged client–solicitor calls. Landowners spotted the spies because the ex-RCMP types sat among grandmothers at community events and ate all the cookies.

An investigation by Alberta's privacy commissioner found that the regulator had violated two sections of the Freedom of Information and Protection of Privacy Act by using private eyes to collect information on private citizens. The commissioner also found that the board's claims that they faced security threats (similar to those made against Jessica Ernst) were entirely bogus. In fact, the commissioner said, investigators were "not necessary for the provision of safe environment." That fall, the Edmonton Journal ran a story entitled "Why Is Alberta's Oilpatch Referee So Reviled?" A historian of the energy regulator answered the question: "I have a sense the board has lost its way."

In 2007, David Swann ran for the leadership of the Alberta Liberal Party and lost. During the race, he visited the Calgary offices of Encana to solicit donations. Swann met with Encana vp Gerard Protti and, during their conversation, mentioned Jessica Ernst and the water issues in Rosebud. Protti stopped talking and looked icily at the politician. "Don't you ever mention that name again in our office," he told Swann. Seven years later, the provincial government would appoint Protti to direct the board's latest incarnation: the Alberta Energy Regulator.

To this day, David Swann considers the province's handling of groundwater contamination from coalbed methane activity as "an astounding case of malpractice, negligence, and malfeasance."

For Jessica, things were about to go from bad to worse.

NINE

Fingerprints and Liabilities

IN THE SPRING of 2006, Alberta Environment delivered several gas samples collected from drinking water to Karlis Muehlenbachs, a geochemist at the University of Alberta. Probably fewer than a hundred scientists and technicians have studied the origins and fingerprints of methane in North America, but Muehlenbachs, a Latvian émigré, was one of the early pioneers. Over the years, the globally recognized expert, author of some two hundred scientific papers, has worked with some of the continent's best fingerprinters, including Dennis Coleman and Martin Schoell. Muehlenbachs, an elfish man with a quick sense of humor, invariably described his work as passive. Industry or government would send him a gas sample—say, from a leaking well—and Muehlenbachs then passed it through his spectrometer. He rarely visited the field: he simply analyzed the fingerprints left behind by straying gases. But in doing so he worked on the front lines of the oil and gas industry's greatest undisclosed liability: leaking wells.

Muehlenbachs was formally introduced to the industry's multi-billion-dollar liability in 1993, when Earl Jensen
of Amoco invited him to join in a multi-company initiative
176   to solve a mystery in heavy oil fields on the border between
Alberta and Saskatchewan, near Lloydminster. Industry had
packed wells close together on the prairie to drain difficult formations of small amounts of dense crude. Many of the wells
were slanted or deviated. Due to gravitational forces, these
J-shaped wells proved difficult to seal with cement. Now half
of the region's 60,000 oil wells were seeping gas.

The methane had infiltrated the soil, suffocated crops,
strayed into other energy wells, and contaminated local
groundwater. The scale of the migration worried Amoco and a
half-dozen other companies, including Husky. Based on available data, the companies calculated that it might cost hundreds
of millions of dollars to fix the leaks. Husky estimated that 47
percent of its assets, or 9,000 wells, bubbled methane. Their
average cost of repair came out to $130,000 a well.

When the EUB threatened to suspend new well licenses
until industry plugged the leaks, the companies formed the
Lloydminster Area Operators Gas Migration Team. The group,
which included Mike Watson, the geologist who would later
help Ernst, commissioned a number of studies. Karlis Muehlenbachs got a specific assignment: to identify the origin of
the leaking gases with isotopic fingerprinting. "At that time the
industry was very open and proactive," recalls Muehlenbachs.
"They weren't hiding anything."

Tracking the origins of methane is a delicate science. Like
criminals in a police lineup, methane flowing from coal,
swamps, or landfills possesses different fingerprints, or carbon
makeups. In the 1970s, scientists divided methane fingerprints

into two genetic camps: bacterial-made (biogenic) and heat-generated (thermogenic). Methane made by microbes such as swamp gas tends to be depleted in a specific carbon isotope, while methane from deep oil and gas formations have a different range of fingerprints. ("My wife still complains that I've never been able to explain what an isotope is," says Muehlenbachs.) Deeper methane also travel with heavier hydrocarbon baggage: ethane, propane, and butane.

But the picture is complex. Lots can happen to methane once it leaves a rock formation. For starters, natural fractures can mix gases at different levels. Bacteria can attack and destroy methane as it rises through the earth, too. "It's very rare that you find a perfect idealized gas," says Muehlenbachs.

Muehlenbachs and his graduate students Kate Rich and Devon Rowe found a way to solve the riddle at Lloydminster. By comparing the fingerprints of gas bubbling around wellheads to various gases found in mud-drilling logs, the geochemist identified an unexpected source. The gas wasn't coming from the targeted oil formation some 1,950 feet below the ground: it came from shale formations halfway down. There, bacteria were eating shale and farting methane. Poor cement jobs—an abiding industry problem—combined with local geology (cement doesn't stick very well to shale) accounted for the leakage. Industry sealed wells so shoddily, in fact, that a 1995 industry study had disclosed that 15 percent of all cement jobs failed, costing the industry nearly half a billion dollars a year.

During the Lloydminster investigation, industry also conducted several studies looking at the impacts of straying methane on groundwater. Nobody from industry or government would likely admit it today, says Muehlenbachs, but in the 1990s methane strayed into groundwater with some regularity. A graduate student working on the Lloydminster research identified as many as a dozen water wells that industry leaks had turned into methane milkshakes. "There was no animosity" about the finding, recalls Muehlenbachs. "The companies took responsibility and simply drilled new water wells." One of the Lloydminster studies even acknowledged that "methane plumes do occur in the groundwater around production wells affected by gas migration." But the highest level of methane found in groundwater was 19.1 milligrams per liter near a leaking energy well. In contrast, methane levels in Rosebud water wells ranged from 30 to 110 mg/l after Encana fracked two hundred shallow wells in the freshwater zones.

Transparency about methane leakage had disappeared by 2006, as industry deployed more force to extract more difficult hydrocarbons from coal and shale. Despite the Lloydminster evidence and data from another oil field, where as many as 80 percent of the wells leaked, Alberta's regulators officially described gas migration into groundwater as "a rare occurrence." In public meetings, they said it never happened. The EUB, which had championed isotopic fingerprinting as a great forensic tool in 1999, based on the Lloydminster studies, now dismissed the technology as untrustworthy. (When Alberta Environment found an ethane fingerprint in Dale Zimmerman's well that directly matched that of a nearby energy formation fracked under Zimmerman's water well, the EUB protested that the ethane sample was too small, and also immaterial because "Zimmerman's well is completed in a coal zone.") The board stopped publicly reporting leaky wells in 2000; its final report on the matter recorded 814 gas migration problems in the province.

This change in attitude baffled Karlis Muehlenbachs. He describes the retreat from truth as a sociopolitical question beyond his skills as a fingerprinter. "Maybe the lawyers got involved and decided that industry can't admit you ever make a mistake," he says. But there is no doubt in his mind that fracking has aggravated a serious existing liability. "You have these horizontal wells that industry now fracks repeatedly twenty or thirty times. In the old days, you fracked once or twice. But each time you frack, it's like banging the well with a giant sledge hammer. It's bound to increase the problem. The critical problem remains the cement and the sealing."

When Muehlenbachs began looking at samples from CBM-contaminated water wells in 2006, he had not only a few government specimens to analyze but water samples collected by landowners, too. Most people no longer trusted the government. "Somehow they found out that I'm doing this analysis and approached me separately," Muehlenbachs recalls. Jessica Ernst was the first to make the journey to Muehlenbach's lab. She brought sampling bags full of gas from a variety of wells, including Fiona Lauridsen's. Muehlenbachs and Schoell later held a two-day workshop on the ABCs of gas fingerprinting for Alberta landowners; Ernst attended.

After reviewing data from the samples, Muehlenbachs reached some vexing conclusions. Neither industry nor government had collected good baseline data on the water quality or the gases found in existing energy wells. As a consequence, the situation was neither black nor white. But with more than 350,000 oil and gas wells drilled in the province, Muehlenbachs no longer regarded the province's aquifers as pristine. "We've been drilling for seventy years today," he says. "There are leaks everywhere." And there was contamination, too.

The fingerprints he analyzed showed methane straying not only from CBM wells but from older, deeper oil and gas facilities already on the landscape.

180   In the case of Zimmerman's well, Muehlenbachs thought the contamination likely resulted from industry activity. The regulator had no good baseline data on methane water content in the region, but Alberta Environment records for the Zimmerman water wells clearly read "Gas Present—No." In the Rosebud area, Muehlenbachs identified propane and butane in several water wells, a clear signature of leaks from deeper gas formations. Microbes, he added, do not make ethane, propane, or butane: "Unless someone threw a Bic lighter down the well, it's a sure sign of contamination." But the lack of good baseline water data again clouded the picture. "What gas was there in the first place and how much was added—you have to guess," Muehlenbachs explains. The available data suggested that leaks from old wells at a range of depths were now mingling with CBM leaks into groundwater. High-volume CBM fracks could also have compromised the casing on older wells.

At the time, Muehlenbachs told *Canadian Business* magazine that the scale of resource exploitation had galloped ahead of the basic science on groundwater. He thought it was shameful that neither industry nor government regulators knew much about the state of groundwater in one of the most heavily drilled landscapes in North America. "They need to have some curiosity about how Mother Nature works and what happens when we fiddle with it," he observed.

On May 8, 2007, Jessica Ernst testified in Ottawa on coalbed methane development before the Standing Committee on Environment and Sustainable Development. She did so on her own dollar, so she could talk freely. David Pryce, an oil lobbyist with the Canadian Association of Petroleum Producers, told the parliamentarians that the "natural gas from coal" held huge economic potential and already represented 8 to 10 percent of the nation's gas supply. "There are technical challenges we're working through," Pryce assured the committee, "and there are strong regulatory rules in place to manage it."

Ernst gave the parliamentarians a radically different version of events. "I have never seen such atrocities in my 25-year career of working in the oil patch as I have now seen in the boom: human rights violations, environmental degradation, and disrespect of the legislation and the regulations," she testified. She showed the committee members pictures of flaming water and explained how Encana had directly fracked into the aquifer in Rosebud. "We are told that only nitrogen is used, so our water is safe because nitrogen comes from the air," she continued. She then showed the committee a list of fracking chemicals from Oilweek magazine. "Some of them contain diesel and mineral oil. In Alberta, the regulator does not require industry to disclose any of the chemicals used, not even if they're toxic, not even if it's benzene, a known carcinogen, or toluene, which damages the brain, notably in children. Toluene was found in our water."

She warned the committee that coalbed methane would expand and also that "the shales are coming. They will spread far. These impacts, violations of the Canadian Charter of Rights and Freedoms, will spread through the country if we continue to allow industry to rule."

One of the parliamentarians said the only time he had seen a toilet on fire was when his kids watched the movie Home Alone: "So it's a surprising and startling thing when you see the photograph that Ms. Ernst just presented to us, showing fire shooting out of her commode, so to speak. It comes across as irrefutable evidence of the problem." When lobbyist David Pryce countered that the gases were "naturally occurring," Ernst replied that having 30 to 66 milligrams per liter of methane in your water was anything but natural. (Later tests by the regulator found methane levels up to 110 mg/l.)

Back in Rosebud, more and more people regarded Ernst's outspokenness as a threat to Encana's promised funding for the Rosebud Theatre. Abusive voicemail messages left on her phone told Ernst to shut up because everyone loved Encana's money. Others called her crazy. One letter to the *Drumheller Valley Times* attacked Ernst in all but name: "I'm tired of people putting down this valley for their personal preferences. If they don't like it leave." In response to the growing nastiness, Ernst pulled back from the community. She didn't like being singled out in the neighborhood, nor being hated for drawing attention to inconvenient truths. She felt a strong chill running through her life.

Although Alberta Environment received nearly a hundred water-contamination complaints between 2004 and 2006, it ultimately investigated only five water wells, including those belonging to Jessica Ernst and the Campbells. In October 2007, alarmed by repeated public protests and Jessica Ernst's vocal criticism, Alberta Environment handed off the job of analyzing the water tests and methane fingerprints to the Alberta Research Council (ARC), billed as a science arm of the Alberta government. Alberta Environment promised an impartial study, advertising ARC as a disinterested and independent party.

ARC, however, was neither. Established in 1921, the research council had initially performed basic science on coal deposits and the tar sands and often in the public interest. But under the leadership of John McDougall, an engineer with close ties to Alberta's ruling Conservative Party, ARC had evolved into something more commercial. Its six hundred scientists now largely did private contract research for corporations. Encana, as it happened, was one of ARC's largest clients. According to the council's 2007 annual report, Encana had funded some of the agency's water investigations, including studies on the impact of CBM drilling. ARC's research hydrogeologist, Alexander Blyth, had contracts with Encana too. ARC billed itself as a private nonprofit organization that did relevant research "for the good of the province," but its website spelled out its corporate sympathies: "We deliver innovative science and technology solutions, meeting the priorities of industry and government, in Alberta and beyond."

ARC wasn't an unbiased player on coalbed methane, either. During the 1990s, the agency had actively promoted and explored the possibilities of cracking coal seams with nitrogen and carbon dioxide (CO_2) fracks. It had also done research on injecting coal seams with microbes to enhance methane production, along with nutrients to feed the bugs. ARC scientists thought that a major solution to climate change could be engineered by storing massive amounts of CO_2 in coal seams.

The council's CO_2 experiments ended dismally: the economics of cooling, liquefying, transporting, and then injecting shallow coal seams with CO_2 were appalling. One ARC researcher, Stefan Bachu, also warned that injecting CO_2 into coal might not be safe. "The coal beds in these regions are relatively shallow (close to 300 m depth), and CO_2 stored in these coals will likely sterilize shallow coal resources that may become economic for mining at some time in the future, while any leaked CO_2 from these coals will likely contaminate groundwater resources in these agricultural regions. In addition, CO_2 storage with CBM production may raise conflicts with land use and public opposition," he wrote.

Alberta Environment's terms of reference for the ARC groundwater contamination review (which were kept secret for years) revealed a troubling institutional bias. The review's mandate was not only to "inform the general public about the findings" but also "to increase public confidence in the Alberta Environment complaint investigation process." To that end, the council had seemingly arrived at a conclusion before it examined the evidence. Four months before ARC signed the contract to review Alberta Environment's contamination data, John McDougall, the council's CEO, provided politicians in Alberta and Ontario with a preview of the results. In a May 2007 briefing memo to provincial cabinet ministers, McDougall offered this assessment: "Rural landowners are anxious for answers to confirm their perceptions that coal-bed methane is responsible for contamination of their water wells. However, results may indicate problems associated with the wells in question may be the result of poor well maintenance or naturally occurring methane." Landowners, McDougall's briefing added, "may not willingly accept the findings determined by Alberta Environment and Alberta Research Council." Jessica Ernst would later obtain McDougall's memo through a freedom of information request. In 2010, Prime Minister Stephen Harper appointed McDougall to the National Research Council, where McDougall promptly cut federal funding for basic science in Canada and championed something called "corporate innovation" instead.

As part of its investigation for Alberta Environment, ARC supervised the drilling of three monitoring wells in and west of Rosebud to get "baseline" data and track contamination. But the data wasn't truly baseline, because the industry had been fracking coals and sands around the community for years. In one Rosebud monitoring well, the researchers found no water, only ethane and methane, which they illegally vented into the atmosphere. Debbie Signer called 911 and forced the government to properly cap the well. In another monitoring well, ARC found arsenic and hexavalent chromium, known carcinogens, but the agency did not alert the community.

In January 2008, the Alberta government released the Alberta Research Council's "Independent Review of Coalbed Methane Related Water Well Complaints Filed with Alberta Environment," with Alexander Blyth cited as author. The review was a thirteen-page summary of water contamination data on the Signer, Ernst, and Lauridsen wells in Rosebud and the Zimmerman well in Wetaskiwin. (The Campbell well was referenced in a separate document.) The conclusions of the summary report didn't surprise Ernst. They reinforced what the government, industry, and regulators had pretended all along. "Energy development projects in the areas most likely have not adversely affected the complainant water wells," said the report. Individual reports were delivered to the landowners in question by uniformed civil servants driving a black SUV and offering the assurance, "There is nothing wrong with your water."

The ARC report blamed everything on either nature or unhygienic landowners. The review found that one water well was contaminated by biogenic methane "naturally found in shallow coal," while the "bacterial/microbial problems" suffered by the rest were likely due to poor well maintenance and "surface water entering the well." Three of the four wells had "coliform bacteria numbers too numerous to count on at least one occasion," the report said. Moreover, water wells completed in coal seams often had gas present, and therefore

methane contamination was to be expected.

Nothing in the ARC report squared with reality or actual science. The historical records for all of the investigated wells clearly stated "Gas Present—No." Ernst knew that fact because she had obtained and checked those records. ARC's attempt to blame well owners for shoddy maintenance confounded the public record, too. Ernst, for example, regularly shock-chlorinated her water well, and her well was safely housed in a barn far from any surface water. As Ernst would later learn through a freedom of information request, notes taken by Alberta Environment and shared with ARC for the review made no mention of Ernst taking care of her well improperly, either. In addition, no E. coli bacteria was ever found in Ernst's well. Only one of the four investigated wells truly showed E. coli bacterial contamination, and that was Debbie Signer's. In Signer's case, Ernst had watched a government employee introduce bacteria into a clean well by mixing the testing rope and water sampler with gopher shit.

Ernst had seen many scams while working in the oil patch, but nothing as bad as the ARC report. The document read like a primer on how to manage industry liabilities. It identified volatile compounds such as tert-Butyl alcohol and phthalates (chemicals found in fracking fluids) in some of the wells, including Ernst's, but dismissed their presence as inconsequential. The review reported the petroleum distillate found in Ernst's water as being at concentrations 1,000 times less than as stated on an earlier independent lab report. The increases in barium, strontium, and chromium found in Ernst's well were not reported in the summary at all. The methane that bubbled into the water wells in question, which ranged from 30 to 110 mg/l, didn't have the same fingerprint as methane randomly found in gas wells tested 100 miles away, the ARC review said; 187 it was therefore natural. Some of the gas might "theoretically" have come from CBM wells, the review acknowledged, but it was an insignificant percentage. The Encana 05-14 CBM well near Ernst's property that had fracked the aquifers made a brief appearance in the individual reports delivered to Signer and Lauridsen: "One CBM well... had perforations and fracturing in the same aquifer that many of the local residential wells are completed in... It is unlikely these short-lived perforations had any measurable effects on the complainant wells at a distance of 1.7 to 3.1 km away." The ARC report omitted the fact that cement squeeze treatments and 18 million liters of frack fluid had been injected into the aquifer, and that examples of frack impacts at greater distances had already been documented.

A cheerful government press release accompanied the ARC report. (Alberta, one of the largest producers of hydrocarbons on the continent, also has the largest public relations department of any government in Canada.) Water-well quality had nothing to do with the drilling and fracking of 10,000 coalbed methane wells, the media release reassured readers. "Many water wells in Alberta are constructed in coal seams that naturally contain methane gas and proper venting of well systems is critical," declared Alberta's environment minister, Rob Renner. (Renner's own department, remember, had reported gas present in only 906 out of 500,000 water wells in the province in 2006, or 0.18 percent of all wells.) What the province needed now, the press release said, was an extensive water-well education program for recalcitrant well owners. The inconvenient fact that thousands of shallow gas wells might connect or fracture into thousands of existing and leaking energy wells in central Alberta, and thereby send methane migrating into water wells, remained an unmentionable truth in a province dependent on hydrocarbon revenue. The minister, a florist by trade, later told a reporter, "It's not the role of Alberta Environment to advocate on behalf of the environment."

The ARC review had not been peer reviewed, and it contradicted several government reports showing that natural gas migration from tight coal seams was rare. A 2006 Energy Resources Conservation Board study on water chemistry in coals, for example, found that most of the water wells in coal seams "had no measurable methane in the water." In fact, 90 percent of the wells in coal showed no detectable methane or ethane. The ARC report also negated the work of Bill Gunter, a prominent ARC scientist. In 2003, Gunter had told a government newsletter, "Since Alberta reservoirs are considered tight there have been very few cases where natural methane leakage has occurred." In other words, there wasn't a lot of methane in groundwater in the region prior to industry's fracking boom.

The ARC report was also marked by glaring omissions. It made no mention of the fracking fluids or high nitrogen levels found in some of the investigated wells. The report rejected isotopic fingerprints supplied by an expert, Karlis Muehlenbachs, on nearby shallow Encana CBM wells. The review avoided all reference to the well-documented complications associated with cracking highly fractured coal seams, such as fracking out of zone into water. No attempt appeared to have been made to check fracking chemicals, water treatment, drilling, or cement additives from Encana or from industry wells fracked above 200 meters. Nor did the ARC report bother to reference rigorous U.S. studies on groundwater contamination, such as the Chafin report. Even the EUB's "high risk" shallow fracking directive got short shrift. Although the ARC report blamed bacteria for the methane water-well contamination, it failed to note that many CBM seams contained industry-sought-after biogenic methane and that fracking companies were actively injecting microbes into coal seams as well as feeding these bugs to increase methane production. Ironically, ARC had sponsored the research.

189

Finally, the ARC review unwittingly confirmed the chaotic nature of water investigations in the province: "Alberta Environment does not have a specific and documented response process with required tasks and decision points to direct the investigative process or the involved parties." Data gathering was "somewhat subjective." Although regulators had licensed more than 300,000 oil and gas wells in the province, the review said, they had no protocol for testing for gases or hydrocarbon pollutants that might be leaking into groundwater.

Although it was not part of the summary review, ARC completed a separate report on the Campbell well in Ponoka. The report found contamination from deep gas but concluded there wasn't enough data to determine whether that was due to a leaking resource well or "a natural fault." In 2013, the Campbells would get a predictable letter from the ERCB. The letter confirmed the presence of hydrocarbons in their well, including high levels of hydrogen sulfide (H_2S, a neurotoxin) and even higher levels of methane and ethane. However, the letter stated, "There is no evidence of a link between energy development activities and the hydrocarbon and H_2S gases present in your water well."

Karlis Muehlenbachs, the expert isotopic fingerprinter, took one look at the 2008 ARC summary report and called it "total bullshit." He and his research associate, Dr. Barbara

Tilley, promptly wrote a two-page critique which raised three key points. For starters, the Alberta Research Council had compared the methane isotopes found in four contaminated wells in CBM country with those found in water wells that were located less than half a mile from intense CBM activity. Given that Muehlenbachs's own lab tests showed the comparison wells "had probably been impacted by CBM activity too," the government was comparing one contaminated pool with another contaminated pool.

Second, said Muehlenbachs and Tilley, the study disregarded key ethane isotope data. Ethane fingerprints make a better tool for diagnosing industry gas contamination, the expert explained, because ethane is largely associated with deeper coal, oil, and gas wells. Ethane is not made by bacteria, and its fingerprint changes with depth, whereas methane's doesn't. The Ernst, Lauridsen, and Signer wells, wrote Muehlenbachs, all had high ethane signatures, indicating that the gas had probably migrated from greater depths in the Horseshoe Canyon formation into their water.

Muehlenbachs's third point was that the ARC study didn't compare coal gas fingerprints from zones of water production with fingerprints from local CBM drilling. The review hadn't even used coal isotope data, which would have been the proper scientific thing to do. Since the "isotope values of ethane in the complainant waters are similar to that of the CBM wells... there must be a contribution of gas from the CBM coals of the lower Horseshoe Canyon formation," Muehlenbachs wrote. The ARC report had dismissed this evidence as immaterial.

Jessica Ernst would later discover through a freedom of information request that the Alberta government was fully aware of the importance of the ethane data. Midway through its protracted investigation, Alberta Environment had changed the lab it used to do the gas analysis. It stopped sending gas samples to Muehlenbachs's lab, which had a reputation for highly sensitive ethane detection capabilities, and switched to a lab at the University of Calgary with much less sensitive ethane detection capacity. The ARC summary had also changed University of Calgary ethane lab results from "not analyzed" to read "not detected." Ernst suspects the government did it all to protect Encana.

191

A year after the ARC study was released, an article on CBM published by Schlumberger, the fracking services company, would disclose what the Alberta Research Council could not: "For a conventional gas reservoir, a fracture stimulation growing out of zone will generally impact only the quality of production. Because of the shallow depth of many CBM basins, the potential exists for a fracture growing out of zone and affecting freshwater aquifers. A thorough understanding of the rock properties can help minimize the possibility of this occurrence." That same year, a study commissioned by Alberta's energy regulators disclosed at a symposium in Tulsa, Oklahoma, that both high-pressured fracking and the injection of acid wastes into abandoned oil fields could have a "negative impact on wellbore integrity." The researchers gave fracking "a high risk score" for damaging wells "due to higher treatment pressures and deeper penetration into the reservoir." Just as Muehlenbachs had warned, the researchers concluded that "the more times a well is stimulated, the higher the chance that casing, cement and or cap rock systems may be damaged."

Landowners rejected the ARC report's findings outright, just as John McDougall's 2007 memo had predicted. One letter published in the *Edmonton Journal* noted that ARC scientists "would have us believe that, as soon as CBM arrived, these land-owners suffered collective amnesia about how to care for their water wells. In spite of the suspiciously coincidental timing, CBM had nothing to do with rendering the water unusable and dramatically dropping static water levels. It's telling that ARC focused entirely on one chemical—methane—which can occur naturally in well water in small amounts. The other chemicals present in all the tested wells, almost certainly from petroleum industry activities, are dismissed. Of course, the council's findings support industry's claims of innocence... Why is Alberta Environment so unconcerned about the likelihood that CBM is poisoning and depleting aquifers?"

In March 2008, eight landowners sent out a press release that publicly disputed the conclusions of the ARC report, as well as the government's decision to stop delivering water to those affected. (The government's earlier commitment to provide water "now and into the future," regardless of "whether the methane was from natural flow or not," had evaporated with the ARC report.) Debbie Signer, who would soon sell her place and remarry, rejected the closure of her case "because of inadequacies, omissions and inaccuracies." The Campbells asked why the government was ignoring data on the discovery of sour gas, a potent brain killer, in their water well. Fiona Lauridsen demanded the public release of data used to make the government's dubious conclusions. (To date, the government has refused to release this data.) Jessica Ernst red-flagged dozens of issues. Encana had originally promised to not frack local aquifers, she wrote, but did so anyway. "In my experience fractured vessels leak. The ARC claims that Encana's perforations were remedied by abandoning the gas well. My question is how does abandoning the gas well repair the fractures?"

That summer, Lisa Bracken, a Colorado landowner, sent Ernst another example of regulatory denial. Bracken, who had read about Ernst in the news, started a website called Journey of the Forsaken shortly after Encana fracked a shale gas well on the uplands above West Divide Creek in Colorado's Garfield County. After the well experienced a blowout in 2004, natural gas and benzene hemorrhaged into the aquifer and onto the surface of a creek barely a mile away. The Colorado regulator reluctantly began a study. Amid continued drilling and fracking, however, a second seep emerged in the same vicinity in 2008. A year later, hydrologist Geoffrey Thyne would conclude that methane levels in water wells and surface ponds were slowly creeping up as the number of energy wells multiplied on the tablelands above. The Colorado Oil and Gas Commission, with the help of Encana, went to great lengths to discredit Thyne's findings. In one memo, the commission dismissed "apparent trends of increasing methane and chloride with increasing natural gas drilling" as not being statistically valid. The commission later ruled that the contamination had been natural from the beginning.

Ernst kept tabs on the denials in Colorado, and she also learned that Encana didn't treat the Canadian military any better than it did landowners with contaminated water. The company applied to puncture and frack shallow rock under a national wildlife reserve on the Suffield military base in southern Alberta, and during federal public hearings on the controversial application, a cache of emails revealed a history of bitter disputes between Encana and military commanders over the company's chronic noncompliance record. In one 2005 memo, former base commander Dan Drew complained that Encana had a singular strategy when it came to breaking the rules: "to deny, delay, deter and deflect." Drew later told *Edmonton Journal* reporter Ed Struzik that he'd rather be hunting down the Taliban in Afghanistan with his son than dealing with Encana on oil and gas issues. Ernst had a good idea what he meant.

TEN

The Police
Come Calling

LONG BEFORE THE Alberta government officially dismissed the methane contamination of water wells as a natural event, Jessica Ernst had decided to sue. The EUB's 2005 banishment letter convinced her of that costly necessity. Someone, she thought, had to hold "these hideous men" accountable. "What are they hiding, and how many people are they treating this way?" she wondered. Originally, she had planned to sue over the compressor noise. Now that she had stumbled upon the cover-up of widespread water contamination, she felt emboldened to go further.

Ernst had no doubts about who she'd have to sue: the two regulators who worked for the Alberta government, one of North America's most powerful petrostates, and Encana, one of the continent's largest shale gas miners. (The company was by then fracking, or preparing to frack, shales in Wyoming, Louisiana, Texas, Colorado, British Columbia, and Michigan.) It seemed equivalent to taking on ExxonMobil and the Texas

Railroad Commission at the same time. However, "if you don't try to do the impossible, the impossible will never be done."

Therapy had taught her that lesson fifteen years earlier, when a psychiatrist at Holy Cross Hospital instructed Ernst to banish from her life the idea of being nice. There was only one way to avoid getting caught as a victim again, and that was to stop being meek and quiet. Most of the landowners she had helped were nice, trusting people. Ernst had abandoned that social costume long ago. On many a job site, workers had called her a bitch. The acronym, Ernst would reply, means "Being in Total Control of Herself."

In her mind, Ernst had carefully previewed the case before her. A powerful company, a pioneer in the fracking of unconventional resource plays, had clearly broken the law. The company had fracked into aquifers at Rosebud with impunity and then diverted water without a permit. Two provincial regulators had failed to investigate the law-breaking. Instead, they had conducted a farcical water investigation that included sampling water from wells with dirty equipment—tainted by gopher shit, no less. The regulators had also failed to follow their own enforcement rules. Rather than penalize Encana for fracking into groundwater, the EUB had attempted to intimidate and punish Ernst for comments she had made about the regulators' incompetence, both publicly and privately. In doing that, the EUB had infringed upon her right to freedom of expression as guaranteed by the Canadian Charter of Rights and Freedoms. The provincial government was covering up the methane contamination of water caused by the industry that had employed Ernst for nearly thirty years. She knew she would need one helluva lawyer.

For Ernst, the time for restitution had arrived. The cycle of abuse she and other landowners had experienced at the hands of regulators had brutally revived memories of her abuse as a child. The parallels did not rest easily with the scientist. The guilty parties were blaming the victims by trying to shame them into obedience and silence, so that more victims could be abused easily. Rapists, pedophiles, and abusive spouses worked that way, too. As much as she wanted to escape the nightmare, Ernst realized she could not walk away from this fight. But she needed a lawyer she could trust.

Ernst began by discreetly interviewing a variety of potential candidates in Alberta. All had some ties to the oil patch. Not one recommended suing the government. There was no money in suing such players, explained the lawyers. "But I'm not doing this for money, and I'll pay you," replied Ernst. Immunity clauses excluded regulators such as the EUB from civil action, the lawyers insisted. Some quoted Section 43 of the Energy Resources Conservation Act by heart: "No action or proceeding may be brought against the Board or a member of the Board or an officer, assessor or employee of the Board in respect of any act or thing done purportedly in pursuance of the Act." But Ernst considered the EUB the most guilty party. What about a claim based on the Charter of Rights and Freedoms? she persisted. The Charter superseded anything provincial. It was there to protect citizens against state bullies. Forget it, the lawyers said. Alberta judges didn't like the Charter, because the petrostate regarded the federal statute as a meddling cook in a kitchen run by corporate rights and money.

Ernst started to think she'd never find a lawyer with the necessary grit. Then, in early October 2007, she decided to call a firm in Toronto: Klippensteins. The firm specialized in difficult legal cases in the public interest, and it had sued negligent governments and corporations. "We are justice-centered, which means that we care both about our clients and about providing the particular solutions they need, and also about doing work which will enhance our communities and make

our common world a better place," read the firm's website.

One of the firm's most famous cases involved Ojibwe activist Dudley George. In 1995, an Ontario police sniper had shot and killed George while the unarmed activist manned a blockade to protect Aboriginal burial grounds and recover lands granted to the Stony Point First Nation. George's brother, Sam, subsequently brought a lawsuit against the Ontario premier, government, and the provincial police. Sam only wanted to expose the truth. The legal battle went on for nine years, but it ultimately forced a major public judicial inquiry. As just one result, the government of Ontario finally created a Ministry of Aboriginal Affairs. In the end, Sam George asked the government to return the treaty lands to their original Native owners. Incredibly, the government did just that. The trajectory of the George case appealed to Jessica Ernst and her Germanic sense of justice.

The Toronto law firm answered Ernst's initial queries with an invitation to take part in what Ernst viewed as a thorough vetting. Klippensteins wanted to find out if the Albertan had what it took to endure a long, hard road. Basil Alexander, one of the firm's lawyers, began the interrogation. Did Ernst have the mettle to stick it out for twelve or more years? he asked. That's how long these cases lasted. Could she sacrifice her life for that long and keep her sanity? Most lawsuits settle in the end because people get worn down and give up, Alexander added. What were Ernst's motives? Was she doing this for money or for revenge? Because the firm didn't want to have anything to do with that. Had there been much media? The volume of newspaper and magazine clippings she was able to produce would surprise Alexander, Ernst recalls.

Did Ernst have the money to pursue this? continued the lawyer with a singular frankness. The case could cost her more than a million dollars even with the firm's discounted rates. Ernst quickly did the math in her head. If she lived ultra frugally, sold the parcel of land she owned in Priddis, and was prepared, if necessary, to sell her art collection, a cabin in Saskatchewan, and her home at Rosebud, and to cash in her investments and Retirement Savings Plan, she'd manage. She told Alexander she didn't have a husband or children to worry about. Unlike most Canadians, she carried no debt. She was free to do battle.

The million dollars, said Alexander next, would cover just legal fees and court costs. And what about the emotional abuse? he asked Ernst. The Alberta government and Encana were powerful adversaries with lots of legal resources, and they would fight back. Was Ernst prepared for that?

Ernst assured Alexander she was ready, and already battle-scarred. She hadn't been able to stop the pedophiles who abused her as a child, but she now commanded the resources and expertise to challenge the professionals who had knowingly raped her community's aquifer. "I don't have gunshot and shrapnel, but a lawsuit is not a bad way to fight a war," she would later tell her lawyers.

After the audition, the firm's principal, Murray Klippenstein, called her. The fifty-five-year-old gravel-voiced barrister rode a bicycle to work and smoked liked a chimney, though not in the office. He had grown up in a Mennonite community in southern Manitoba. He viewed the workings of courts not just as a technical exercise, but also as a profound historical narrative that sometimes delivered justice. The law moved like a glacier, and like a glacier, he said, it exerted great powers.

Klippenstein asked Ernst if she would settle out of court if Encana offered her millions of dollars. Ernst said no fucking way: "Murray, I'm not doing it for money. I'm doing it for truth and justice." She explained that she wanted to expose what had happened, on the public record. There would be no cash settlement wrapped in gag orders and no sealing of court records. "We saw eye to eye on that right away," Klippenstein recalls.

Ernst then turned the tables and asked Klippenstein a question. Why did he choose to study and practice law? "You can practice law to serve the devil or serve your soul. I'd like to believe I'm the kind of lawyer that practices law to serve the soul."

Did he have the guts to take on a petrostate and its key corporate client? "What would you do, Murray, if Encana offered you $10 million to get rid of me?" she said.

Klippenstein told Ernst a story about the Ipperwash case. Nearly eight years into the lawsuit, some government lawyers suggested a meeting at the firm's boardroom, where they offered a large amount of money for the George family and millions in legal fees to Klippenstein if they settled out of court. Klippenstein said no in a heartbeat. After the offer, he told Ernst, his hands shook uncontrollably for a half an hour. Both Sam and his firm had exhausted their finances long ago and faced a six-month trial. Klippenstein's family car was a wreck: to drive it, the lawyer had to climb in through the passenger window, because the door wouldn't open. But Klippenstein told Ernst he had never regretted their decision. He went on to win the case, too. (Sam George died of cancer in 2009 at the age of fifty-six.)

During the series of phone conversations that followed, Klippenstein talked to Ernst about the things that killed law-suits like hers. The first was science. Only the law could truly win a lawsuit, because that was something judges understood. In Ernst's case, Klippenstein said, it looked like Encana and the government had broken several laws.

The second thing that killed lawsuits were the plaintiffs themselves, said Klippenstein. The procedural wrangling alone often took years. Before even reaching trial, which would be grueling and nasty, many plaintiffs found the lawsuit to be a road to disappointment, and bitterness overwhelmed them. Ernst replied that sacrifice came to her naturally. Moreover, she said, she wasn't doing this for herself. Apart from love, only three things really mattered in this world, she told Klippenstein. One was groundwater. The second was the integrity of the democracy she lived in; and the third was breaking the cycle of abuse. Klippenstein liked what he heard. "It makes me feel humble to work with people like that," he later reflected. Ernst had found her lawyer.

The final detail Klippenstein raised jolted Ernst. Under court rules, she could win the case but still come out "in the negative," he explained. Plaintiffs bore some staggering risks. "As your lawyer, I have to tell you this. Even if you win legally, because you won't settle, you may have to pay the legal fees and the costs of the defendants' lawyers," declared Klippenstein. "If it looks like you're winning, and the judge starts adding up some prospective settlement, the defendants could phone and say, 'We have an offer of $1.5 million.' If we reject that offer, and the judge finally awards something less than the offer made by the defendants, then you'd have to pay the defendant's legal fees." Ernst sat shocked at the other end of the line. Then she recovered and told Klippenstein, "Well, I

have a plan in place." Klippenstein knew then that he was dealing with a client prepared to go the distance.

During the back-and-forth with Klippenstein, Ernst continued to speak up. In a letter to the *Drumheller Valley Times*, she
commended a decision by the Wheatland County Council to
deny approval for any more gas wells within 1.5 kilometers of
Rosebud. "Considering that Encana perforated and fractured
our drinking water aquifers without conducting any appropriate data collection first... I am pleased to see our council stand
up to the rogue company," she wrote. The editor thanked Ernst
in writing for her letter, saying, "Encana is slowly buying off
people and it's not right. Keep up the good work."

On October 30, Ernst noticed that her dog Magic was
behaving strangely outside. Bandit, Magic's larger-than-life
brother, was nowhere to be found, and that was not normal.
Ernst began to search her acreage with an air of foreboding. By
the railway tracks south of her property, she spotted something
white. As Magic backed away, Ernst came upon a headless dog.
When she picked Bandit up, she thought she was going to die.
She took off her favorite multicolored sweater and wrapped
the mangled dog in it. As she walked home, she screamed. A
farming neighbor, Peter Lauridsen, Fiona's husband, later
helped her find Bandit's head and dig a grave in the flower
garden by the driveway. Ernst's dogs had safely navigated the
tracks for six years.

Everyone in Rosebud had known Bandit, the charming prancer. His gruesome death shocked even those who
believed the government rumor mill that Ernst was crazy
and that Encana would never break the law. Years later, Ernst
would come upon an uncanny newspaper account about
an Australian farmer protesting fracking in her community.
The Australian woman, too, had found her Jack Russell dead.

Ernst, who believed Bandit's death was an act of intimidation, stopped walking in the community, and no longer felt comfortable swinging in her hammock in the backyard.

To manage the pain of her loss, Ernst dove into prepa- rations for the lawsuit. She also continued her scientific investigation on the migrations of methane. Everywhere she looked, gas migration and fracking seemed to go hand in hand. She read about an explosion that had ripped apart a house in Bainbridge Township in Geauga County, Ohio, where industry was fracking shales. Investigators later found that methane had invaded nineteen homes in the township through their water wells. The state's Department of National Resources blamed a 3,000-foot-deep gas well that had been badly cased, fracked, and then shut in. Regulators pegged "hydro fracturing" as a major contributor to the contamination, because the fracking had rattled an already compromised wellbore. With great unease, Ernst read that the authorities in Geauga County "disconnected 26 water wells, purged gas from domestic plumbing/heater systems, installed vents on six water wells, plugged abandoned in-house water wells, plumbed 26 houses to temporary water supplies, provided 49 in-house methane monitoring systems for homeowner installation, and began to provide bottled drinking water to 48 residences upon request."

The shale revolution, which had now invaded Texas and North Dakota, was also repeating the experience of coal and fracking out of zone. In the Bakken Shale, oil companies fracked into each other's wells, shallow formations, and aquifers. "Several operators have reported difficulty keeping fractures contained within the target Bakken horizon," reported one 2007 Society of Petroleum Engineers paper.

Over that fall, Ernst's lawyers assembled the frame of a case. Klippenstein began by checking and double-checking Alberta's statutes and laws. Since the EUB's "banishment" letter had been

received by Ernst on December 3, 2005, her lawyers had to file their claim no later than December 3, 2007. Ernst bombarded Klippensteins with emails full of data, regulatory correspondence, and industry files. There were few peer-reviewed papers on the environmental impacts of fracking at the time, but Ernst sent those too. Despite her conviction that she was doing the right thing, she dreaded what lay ahead. Government and industry had treated her terribly when she was just an ordinary oil-patch consultant and landowner. "What will they do to me when they find out I am suing them?" she asked herself. And what would the residents of Rosebud think?

The first claim, only nineteen pages long, set out the legal basics with more amendments and details to come. It asked for $100 million in damages and noted that Ernst had "25 to 30 times the level [of methane] reported as an explosion risk by the Canadian Association of Petroleum Producers." On the day of the court filing, Klippenstein advised Ernst to take the next six months to think about the implications of the case. If she wanted to quit, they would completely understand. She retreated to her cabin in Saskatchewan and mourned for the animal that had reintroduced love into her life. Four months later, she informed Klippenstein that she was going forward: Bandit didn't like quitters.

Cory Wanless, a twenty-six-year-old Alberta-born lawyer, joined Klippenstein's firm after the preliminary claim was filed. One of his first jobs was to review the tape of Ernst's meeting with Rick McKee. The contents of the tense conversation astounded Wanless. He chuckled when he learned Ernst had asked McKee's permission to record the meeting. That meant they could use it in court.

During her ongoing gas migration research, Ernst had come across the proceedings of the fourth Wellbore Integrity Congress in Paris. Experts from around the world gathered every year to talk about the millions of leaky oil and gas wells, she learned. The second Wellbore Integrity Congress, in 2006, for example, had warned, "There is clearly a problem with well bore integrity in existing oil and gas production wells worldwide." What particularly piqued Ernst's interest was a presentation to the congress given by Theresa Watson, a member of the Energy and Utilities Board. The data Watson presented showed that oil and gas wells leaked from both shallow and deep zones—everything the ARC report denied. Shallow leaks were generally due to "poor cementing practices," Watson told her audience, while deeper leaks were related to fracking, perforating, and acidizing, which could create pathways in cement seals. Watson reported deep leakage to surface and groundwater in central Alberta, even showing a picture of methane bubbling up in a puddle of water in a farmer's field. As industry increased well density from 350,000 to 900,000 in the province by 2056, Watson said, gas migration would get much worse. In fact, Watson speculated, energy wells would collide with water wells so dramatically that gas would start accumulating in the basements of homes "due to migration from shallow zones." Ernst wondered why experts in Paris got to hear this information but not Albertans.

Ernst's lawyers continued to flesh out a case of negligence against Encana and the two Alberta regulators, as well as a Charter claim against the EUB for violating Ernst's right to freedom of expression. In the end, Klippenstein asked for $33

205

million in damages in the claim, to illustrate how badly Encana and the Alberta government had behaved. In December 2008, a year after the initial filing that preserved Ernst's statute of limitations, Klippensteins served the papers on the defendants, who included Neil McCrank, the chairman of the regulator; the EUB's Jim Reid, who had "inappropriately characterized Ms Ernst as having made threats of violence"; and Kevin Pilger, the Alberta Environment water investigator. (The three men were later dropped from the lawsuit due to cost, legal precedent, and pragmatic concerns about simplifying the claim.) The server of the papers told Klippenstein that Jim Reid was so shaken up at the time that he had trouble standing. Reid also blurted to the server that he couldn't be sued. A lawyer from Encana immediately called Klippenstein and asked, "Where are you going with this?"

Klippenstein had a short answer: "I'm going to court."

In late January 2009, Ernst's lawyers flew out from Toronto to tour the fracked wells around Rosebud and to meet their new client. Ernst was anxious. She had written to a friend about the lawsuit, "Never in my life did I expect to be doing such a thing. Needless to say, the butterflies in my tummy are massive and violent." When she met them at the airport, Klippenstein was wearing a cheap and silly cowboy hat and Wanless looked endearingly disheveled. Ernst, used to a certain oil-patch crispness and tidiness of dress, briefly thought, "What kind of lawyers have I hired?"

Ernst drove the two men to Rosebud, where she showed them the 05-14 well, the compressors, and the neighboring community of Redland. Back at her home, Ernst nervously heated up her well room in the barn and reconnected her water well to the house taps. (She was now hauling water every six weeks.) Klippenstein wanted to see the methane problem as Ernst had lived it. The lawyers stood dumbfounded as the methane whistled from the taps like distant trains. Ernst filled a one-gallon plastic milk jug and capped it. She let Klippenstein and Wanless take turns lighting it with a whoosh. Each time the gases pouring from the jug burst into flame, the lawyers shook their heads in disbelief. "They were mesmerized," says Ernst. "They couldn't believe it was true."

On February 9, two months after Ernst's lawyers had served the legal papers, Ernst got a call from the RCMP. When Detective Dennis De Franceschi introduced himself as a member of the National Security Enforcement Team, Ernst started laughing. De Franceschi, who was based in Calgary, explained that he belonged to an antiterrorism squad. Ernst laughed harder.

"You probably watched the national news two nights ago," Ernst said to De Franceschi, "and that's why you're calling." CTV's W5 program had presented a documentary called "Fuelling Fears" about how badly the ERCB was treating landowners in Alberta's oil patch. Ernst's contaminated-well story had appeared front and center. In the documentary, Ernst recounted how the EUB had accused her of making criminal threats yet there had been no police investigation. Now the RCMP had finally called.

De Franceschi claimed he hadn't seen the documentary and informed Ernst that the matter at hand was serious. He had heard that she was good with landowners and that they trusted her, he said. Would she be willing to spend some time training forty-three members of the RCMP's antiterrorism squad?

Ernst smelled a rat larger than a Drumheller dinosaur.

Toward the end of their hour-long conversation, the RCMP detective finally admitted that he was investigating a string of pipeline bombings in northeastern British Columbia, six hundred miles away. The bombers had targeted an Encana pipeline and wellhead near Pouce Coupe after an aggressive shale gas drilling program had unsettled the community. A significant percentage of the gas in the area contained hydrogen sulfide, a brain killer and public health hazard. A note from the bombers not only accused Encana of "endangering our families" but slyly added that the bombers would not negotiate "with terrorists." The first bombing had occurred the previous October, and the fourth and most recent in January 2009. The National Security Enforcement Team, which had put 250 officers on the unsolved case, wanted to send over a two-man team to talk to Ernst about the sabotage. Many farmers around Pouce Coupe had already been interrogated by the police.

Ernst told De Fransceshi she knew nothing about the bombers. She emphasized that he was not to come to her home, either. She and her lawyers would gladly come to his offices, she said, if the RCMP paid the travel bills.

The conversation left Ernst shaking. No one really knew about her legal case yet except the defendants: Encana and the Alberta government. (To protect her case, she would not go public until 2011.) Now she feared for the safety of her legal files, which filled her basement. She quickly backed up her computers and took her external drive to a neighbor's house, concealed in a box of clothes—"just like the French Resistance." Ernst called her lawyer, Murray Klippenstein, and together they plotted strategy. Ernst suspected that the National Security police would call back or, worse, show up unannounced. Were they trying to intimidate her into dropping the case, or did they somehow want to blame her for the Encana bombings? She suspected that Encana had given the police a list of people to interrogate and that her name had appeared at the top.

In an email to De Franceschi the next day, Ernst asked why the police wanted to talk to her. "In my view," she wrote, "it almost seems like the police are coming to me because over the years I have publicly called for accountability of the gas company involved, and embarrassed them, and now is there a chance for the company to even the score and intimidate me by giving my name to the police and suggesting you check me out."

Three days later, Ernst awoke with a bad feeling. She called her neighbors Ken and Theresa Wise, and asked them to come over to possibly serve as witnesses. The Wises arrived in ten minutes. The three chatted and visited; Theresa had brought along her quilting. Within hours, an unmarked white car approached Ernst's driveway. Ernst took photos from the living-room window, then rushed out the door and snapped more photos of De Franceschi and Corporal Dave Bibeau as they emerged from the vehicle and walked toward her. De Franceschi told Ernst she was not allowed to take their photos because they were undercover. Ernst replied that they were trespassing and they needed to call her lawyer. She insisted the policemen use her phone, because she wanted a record of the call.

"Do you have a warrant?" asked Klippenstein from Toronto.

No, said De Franceschi.

"Then you have no right to be there. Please leave."

As the security team prepared to get back into their car, Ernst disobeyed Klippenstein's final instructions to her not to engage the police. Acting on instinct, she invited the men onto her porch and said she would answer their questions. She didn't want the RCMP to think she was hiding anything or to give them an excuse to come back and take her files.

It was cold, and the men asked if they could come inside. Ernst said no. She suspected they wanted to canvass her home. The scientist calmly invited the RCMP officers onto her porch, asked Ken Wise to join them as a witness, and gave De Franceschi a metal chair. She offered the policemen blankets and tea with honey. They declined.

Ernst requested that De Franceshi record the meeting, which he did. The antiterrorist investigator had only six questions. Most of the questions didn't relate to her case, but Ernst recalls that one question stood out as particularly "vile."

Would Ernst hand over the names and phone numbers of all the landowners she had helped over the years, De Franceshi asked.

The number easily exceeded several hundred ranchers and farmers. Ernst bristled: "Over my dead body. I don't betray people."

Ernst asked the policemen who had sent them and how they had got her phone number and address. The cops declined to answer. She asked if they had interrogated Encana or the regulators. No answer again. Ernst and the two officers talked briefly about water contamination and the impacts of the industry on the land. They informed Ernst that they had a job to protect "critical infrastructure" and "visiting dignitaries." Although Ernst wanted to add, "But not Canadian citizens?" she bit her lip.

De Franceshi had one last question: If Ernst were the fire chief at Pouce Coupe, near where the bombings had taken place, what would she do to keep citizens safe?

"That's easy," replied Ernst. "I'd secretly have industry shut down all their sour gas wells and pipelines. I'd put everything in lockdown except for the sweet gas facilities. You then let the bomber do what he wants, knowing that a release of sweet gas won't kill anyone. That way you'll catch the bomber and keep the community safe. It's very simple."

After Corporeal Bibeau had apologized profusely for the intrusion, Ernst says, the National Security investigators left the premises like sheep that had just been shorn. Ernst barely slept for weeks afterward. If the contamination of her water was natural, why would the government send the Royal Canadian Mounted Police to her door?

ELEVEN

Kafka's Law

THE FAMOUS WRITER Franz Kafka, the son of a German-speaking Jewish family from Prague, composed an odd legal parable on the eve of World War I. The story concerns a peasant from the country and his attempt to find justice. The man approaches a doorway that might lead him there and asks a bearded doorkeeper "for admittance to the Law." The stern gatekeeper, a Tartar, says no. Taken aback, the peasant asks, when might he gain entry to the law? The gatekeeper replies that admittance will be possible at some time, but not at the moment. He explains to the peasant that there are more doors down the hall and more keepers but that each is more terrifying than the last. The man thinks for a moment and then decides to wait. The gatekeeper gives him a stool to sit on. The man and the gatekeeper exchange a few indifferent pleasantries. Years pass.

By the end of the story, the peasant has grown old waiting for admission to the law. Finally, near death, he asks one last question. "Everyone strives after the law," says the man, "so how is it that in these many years no one except me has requested entry?" The gatekeeper, noting the man's fragility, answers briefly: "Here no one else can gain entry, since this entrance was assigned only to you. I'm going now to close it."

From the time she filed her original statement of claim in December 2007, Jessica Ernst encountered one gatekeeper after another. Some refused to divulge public information in a timely fashion; others simply consumed time and money. The legal gatekeepers proved to be the most officious of all. Five years would disappear before a judge heard a single argument about the case, due to what her lawyers said were normal complications and paper shuffling for a mammoth legal case. While Ernst frantically waited and fretted, she asked herself, "What kind of fucking life is this?"

The delays bore direct testament to the power of Ernst's lawsuit. Her final seventy-three-page statement of claim pulled no punches: it accused Alberta Environment, the ERCB, and Encana of negligent, careless, and reckless failure to obey the law. In particular, it charged Encana with conducting a risky and experimental shallow CBM drilling program that had fractured an aquifer and released "a large amount of methane and other contaminants into underground freshwater supplies." The suit also accused the gas giant of willfully breaching nearly half a dozen laws, including Alberta's Water Act and its Oil and Gas Conservation Act.

On the government side, Ernst's lawsuit detailed the EUB's repeated failings as a regulator. It accused the board of failing to take reasonable and adequate steps to protect citizens near oil and gas activities from water contamination. The suit said the board, which had described shallow fracking as "high risk," not only owed Ernst a clear duty of care to prevent harm to groundwater but also had failed to uphold its own legislative obligations to enforce the law and to protect the public interest: "The EUB breached the above duties by negligently granting licenses to EnCana to drill shallow CBM wells in the Rosebud area despite the existence of significant risks that drilling these CBM wells would contaminate groundwater." Moreover, the suit claimed, the EUB had violated Ernst's Charter right to freedom of expression by banning all communication with her and "removing her from the standard regulatory process."

Finally, the claim detailed Alberta Environment's "negligent administration of a regulatory regime." The regulator had not only breached the Water Act and the Environmental Protection and Enhancement Act but had "refused to perform adequate testing on suspected problematic hydrocarbon wells" and had "falsified, manipulated, ignored and withheld data." The actions of the government regulators and Encana, the claim concluded, "amount to high-handed, malicious and oppressive behaviour" that justified punitive as well as general damages in the neighborhood of $33 million.

Once the claim was filed, the gatekeepers immediately got to work. The first—and the "nastiest," as Ernst later put it—was Alberta's Tory government. Having ruled the province for thirty-six years, the hydrocarbon-infused Conservative Party had a reputation for operating the most secretive and least transparent government in Canada. A 2007 audit of freedom of information requests in the country by the Canadian Newspaper Association rated Alberta as one of the nation's worst offenders. Alberta's freedom of information process, said the association, worked as a "denial of information" instead.

Ernst discovered the veracity of that damning audit for herself. Months before she hired Klippensteins in 2007, she submitted a freedom of information request to Alberta Environment and the ERCB. That was followed by a request in 2008 to the Alberta Research Council, asking for all the "public" baseline water data mandated for shallow CBM in 2006, as well as notes, references, documents, correspondence, and records on the botched methane contamination investigation. Ernst suspected there might be something useful for her lawsuit in that data, and her instincts proved correct in ways she could never have imagined.

At first, Alberta Environment fought her request tooth and nail. It withheld records and censored legal addresses of both water and gas wells. Ernst complained to the Office of the Information and Privacy Commissioner of Alberta, explaining that she worked in the oil patch and knew that energy well locations weren't private data. After some prodding from the commissioner, Alberta Environment reluctantly released some uncensored records. The documents showed the government knew in 2006 that the ethane, propane, and butane fingerprints of the gas in some Rosebud water wells matched those from Encana's shallow gas wells in the area. Ernst was shocked. That meant the government had known about the liabilities of shallow fracking and that the ARC review was a cover-up for both industry and the regulators.

When Ernst applied again through freedom of information for the specific data used by ARC to dismiss groundwater contamination in the province as a naturally occurring event, she faced more stonewalling. Alberta Environment said it didn't have the records and that Ernst should go to the Alberta Research Council. When approached, though, the gatekeepers at ARC told her to make her request to Alberta Environment, because ARC had given all the records back: "As explained we are unable to provide you with the information." For months the two agencies played Ernst like a ping-pong ball. When ARC finally agreed to answer her request, the council explained it would cost $4,125 to copy and retrieve the material she wanted, with a $2,000 deposit. It sounded like extortion to Ernst, but she complied just to see what would happen.

The first batch of material ARC mailed to Ernst appeared to be totally unrelated to her request. The next shipment was so heavily redacted that the documents were unreadable. The material was also so badly mixed up that it looked to Ernst as if someone had taken thousands of pages and thrown them in the air before boxing them up. As had Alberta Environment, ARC had blacked out not only water and gas well legal land descriptions (routine public data) but also all information on landowners with contaminated water. ARC had even censored the data on Ernst's own water well. The council withheld hundreds of documents on the grounds that their disclosure might be harmful to business or economic interests.

Ernst sifted through the six thousand jumbled pages eleven times. She even created her own autopsy room to piece together parts of the missing puzzle. Brief phrases and obscure references hinted that ARC was hiding incriminating information, including evidence that Encana had provided ARC with gas-well data and that the final ARC report had been heavily edited by the government. Ernst penned more complaints to the Office of the Information and Privacy Commissioner. The office reminded ARC that documents on drinking-water contamination were in the public interest and not to be redacted. When the council ignored the commissioner's advice, Ernst demanded a fee waiver from ARC for "the confusing mish mash of all kinds of stuff."

Her bold request ("I can be annoyingly persistent," admits Ernst) prompted a personal reply from John McDougall, president and CEO of the Alberta Research Council, in December 2008. That wasn't normal protocol. Every public body had a specified individual who managed freedom of information requests, and it wasn't the CEO. But now McDougall—the man who had, even before the agency analyzed Alberta Environment's data, predicted that Albertans would not like the council's findings on methane contamination—was telling Ernst that he was "respectfully" declining her request, since "it has not been reasonably demonstrated that your reasons justify a fee waiver." When Ernst received the missive, she cursed the air blue.

As other freedom of information requests to Alberta Environment and the ERCB floundered, Ernst felt doors slamming on her. Why was it so hard to obtain what the government openly advertised as public information: the baseline data on water and gas wells that ARC claimed to have used to close the province's investigations of methane contamination of groundwater? Ernst realized that she needed to find another door to the information, and quickly.

Ernst asked her lawyers if she should request a formal inquiry by the province's privacy commissioner into ARC's evasion. Klippenstein said that he doubted she'd get results and counseled her to wait for the document-sharing phase of her lawsuit. Ernst disagreed and forged ahead, spending hundreds of dollars and several weeks preparing the appropriate documents to press for a formal inquiry by the OIPC. She would wait a year and a half for a ruling, and the results would later floor Kippenstein.

In the meantime, Ernst ran up against another group of gatekeepers. Four parties had inserted themselves into Ernst's freedom of information requests as "directly affected parties." Without her knowledge, they had been given the opportunity to secretly comment on Ernst's requests for data. Ernst learned about the invisible censors only when the OIP Commissioner casually dropped mention of the parties in correspondence with her.

The secret parties were two major defendants in her lawsuit— the ERCB and Encana—and two major industry players: the mega-fracking company Schlumberger and PetroBakken, an oil and gas company with many fracking interests. "I was never consulted by the Office of Information and Privacy about having invited secret participants in my inquiry about contamination of a public resource—water," wrote Ernst to the OIPC. "The bias against Albertans doesn't get more obvious." These secret parties "should have nothing to hide if nature truly is guilty" of widespread water contamination during shallow fracking, she added.

To support Ernst's fight for the release of public data, several landowners sent letters to the Office of the Information and Privacy Commissioner decrying the way the Alberta government had treated her information request. "How much time, money, anguish, abuse (the list goes on and on) does Ms. Ernst have to endure to obtain public data?" wrote Debbie Signer. "I believe that so much dirt has been swept under the rugs in Alberta that we now have rug piles that rival the Rockies!!"

As Alberta's gatekeepers blocked Ernst's attempts to secure data, life delivered to Ernst's doorstep a string of misfortunes. On December 23, 2009, her nephew Derek, the young man who had loved to wander by the Rosebud River and hear the chorus frogs sing, hanged himself with a rope. He left a hand-written note explaining that he could no longer beat his own demons and thanked "Aunt Jess" for always being there for him. Ernst cried for days and grieved for months. To make matters worse, Ernst's relatives blamed her preoccupation with the lawsuit as a major factor in Derek's death. In April 2010, Gerda Spencer, Ernst's mother, died at the age of seventy-eight, never having made amends for not protecting her daughter from the pedophile who preyed on her as a child. Ernst didn't attend her mother's funeral; she thought it best that she stay away. Shortly afterward, Mike Watson, Ernst's oil-patch confidant and mentor, died of a heart attack.

To ease her grief, Ernst plunged herself once again into data. One day she visited the Alberta Government water-well database to look up the historic records on some of the contamination cases in the province. To her dismay, not only were historic records removed, they were altered. Documents that previously said "Gas Present: No" now offered only blank lines with no information. She frantically backed up all her files.

The fracking industry, meanwhile, worked hard to soften growing public resistance. On its company website, Halliburton compared fracking to child's play: "Sand, water and pressure: the basic components of building a great sandcastle, and the same ones being used today to spur a revolution in the way Americans access and utilize clean-burning energy resources." Talisman, a Calgary-based company, took the theme further by commissioning a twenty-four-page coloring book for distribution in rural Pennsylvania. The comic depicted the adventures of Terry, a friendly Fracosaurus, who told readers, "I am here to teach you about a clean energy source called Natural Gas." Terry efficiently liberated methane from smiling underground rocks for a better America; even migrating gases appeared as happy-faced balloons. After the Fracosaurus visited one community, eagles and rainbows blessed the landscape.

Comedian Stephen Colbert pilloried the propaganda as the fracking industry's equivalent of Joe Camel. Colbert compared fracking to "giving the earth an Alka-Seltzer, if the Alka-Seltzer shattered your internal organs so that oil companies could harvest your juices." The coloring book, suggested Colbert, left out the real story about Terry the Fracosaurus: depressed by the drilling and fracking of his sacred fossil ancestors, Terry probably blew himself up in the shower after methane contaminated his water well.

Like Colbert, many people weren't buying industry's assurances that fracking was safe. Andrew Gould, chief executive of Schlumberger, openly acknowledged in 2010 that there were still many unknowns. "I don't think that the actual optimum technology set for producing shale gas has yet been defined— at the moment, we are doing it by brute force and ignorance," Gould told the *New York Times*.

GasLand, Josh Fox's documentary on fracking, created an uproar when it was released in 2010. The film showed landowners setting their tap water on fire in Pennsylvania and Wyoming. The fracking industry quickly commissioned its own documentary, called *TruthLand*. But in a little-read American 2011 report called "Plugging and Abandoning Oil and Gas Wells," the National Petroleum Council admitted that fracking shale formations had collided with an older problem: abandoned and leaky wells. As fracking placed more wells on the landscape, more gas would leak "into other formations and fresh water," the report said. Not only had industry not done the proper research on materials for sealing these leaky wellbores, according to the report, it lacked "long-term vision" about the multi-billion-dollar liability.

Ernst's stresses continued to multiply like fractures in the continent's shale rocks. Some of her friends, weary of the weight of the lawsuit, asked when it might end. Ernst wondered the same thing herself, but she realized that as soon as she went public with the lawsuit, filing successful freedom of information requests would get harder, if not impossible. Her industry contracts dwindled, and when her last contract in the oil patch formally ended in 2011, her business collapsed. Since engaging the services of the ever-blunt Ernst had become a political liability in the industry, Ernst was now living on her savings.

Murray Klippenstein had warned Ernst that lawsuits proceeded at a snail's pace in their early years, but she was not prepared for the tenacity of the gatekeepers. Another almost fatal obstacle proved to be her own stubbornness. "I'm a control freak, and that can be completely annoying," Ernst warned her lawyers at the outset. But as Klippenstein and Wanless made more demands on her time and as her expenses mounted, Ernst wondered if they had her best interests at heart. Tensions finally exploded over the design of a website that would update the public on her lawsuit. Her lawyers wanted to control the site to protect their client. Ernst thought that was an inefficient use of their time and said so. She also worried that she was losing power over her own lawsuit. On March 21, 2011, an exasperated Klippenstein respectfully resigned from the case with the words "I am sorry that we did not meet your expectations." As the demanding plaintiff read his email, her heart sank.

That weekend, Ernst and Magic went for long walks in the cold. While trudging through the snow, Ernst prayed for clarity on whether she should drop the case. She prayed to God, too, for a sign one way or another. She found that sign classified as "spam" in her inbox: a totally unexpected invitation to speak about the dangers of fracking at the nineteenth session of the United Nations' Commission on Sustainable Development in May. She recalled a Japanese proverb: "Fall down seven times, stand up eight."

On March 23, she penned Klippenstein a short note, writing it on a hand-drawn map showing how Encana had perforated and fracked the Rosebud aquifer in seven zones. "Hi Murray. This case is too important to quit. If you really want a website, let's figure out a way to resolve this and get going forward. We've both put too much time, energy, heart and $ into this to give up." It was signed, "Please, Jess." Within days the team was back on course. Ernst agreed to share work on the website with her lawyers. The site would later record hundreds of hits a day, with volunteers offering to translate Ernst's posts into Polish and French.

A month later, Ernst's lawyers flew to Calgary. In a press conference held at the Kensington Inn, they went public with Ernst's amended seventy-three-page lawsuit. The case had universal importance, Murray Klippenstein said, because "the hydraulic fracturing issues that Jessica raises in her lawsuit" were coming up more and more all across North America. Sitting between her lawyers, Ernst told a room full of reporters, "I'm doing this case for all Albertans, and for our water. I'm not doing this for me." Following the press conference, a CBC reporter asked Ernst, "They have all the power, they have all the money, they have all the courts, and you have nothing.

How can you do this?" Ernst gave a short answer: "I have one thing they do not. It's more powerful than anything: heart."

An Alberta geological engineer who read Ernst's claim on her website wrote a lengthy letter to wish her good luck. "The very high methane concentrations measured are simply irrefutable evidence of groundwater contamination," he observed. "[The] Gas company's only defense is 'prove it' which they know is pretty much impossible without baseline sampling and the inherent uncertainty associated with fracture networks (both natural and fracked)." Another industry employee told Ernst her lawsuit was the best thing ever to happen to the province.

Nevertheless, gatekeepers continued to put roadblocks in Ernst's way. One of the most powerful came in the guise of the federal government under Stephen Harper, who was first elected as prime minister in 2006. Even *The Economist* magazine described Harper, Canada's autocratic and highly secretive leader, as a "political predator" with little regard for science, due process, or the media. As the son of an Imperial Oil executive and an admirer of U.S. oil-funded Republicanism, Harper introduced an air of menace into Canadian politics. He championed oil just as he praised God and free market economics. The prime minister also counted Gwyn Morgan, Encana's former CEO and an outspoken fracking advocate, as one of his government's key political supporters.

Three characteristics of the Harper government particularly disturbed Ernst. For starters, it openly demonized Aboriginal people and ordinary people concerned about unconventional resource extraction. Under Harper, the federal government axed critical environmental monitoring programs and went on to gut some of the nation's oldest environmental laws, all to advance hydrocarbon developments. Among other things, the changes allowed the fracking industry to harm fish-bearing waters with impunity. In fact, members of the Harper government praised hydraulic fracturing as an "amazing North American technology." One minister even suggested that Russia, home to a third of the world's fracking activities, had secretly funded the North American anti-fracking movement.

Harper's overtly ideological government had little regard for evidence of any kind, and that too alarmed Ernst. The federal Conservatives not only muzzled scientific studies on climate change but dismantled the nation's science libraries. They obstructed freedom of information requests as religiously as the Alberta government did. Harper also changed the nation's approach to basic science by appointing John McDougall as the chief of the National Research Council in 2010. McDougall acted in his new capacity as he had done at ARC: he transformed the ninety-seven-year-old government research agency into a contract research organization focused, as the Globe and Mail put it, "on a clutch of large-scale, business-driven research projects at the expense of basic science."

Most worrisome of all for Ernst was that Harper had little regard for the nation's Charter of Rights and Freedoms. His administration openly viewed the Charter as an inconvenient document that limited the power of corporations, along with the reach of the Prime Minister's Office. Harper not only questioned the authority of the judiciary (he thought politicians should make laws, because judges were notoriously too liberal) but attacked the impartiality of the Supreme Court.

On May 2, 2011, the night Harper's Tories won a majority government, Ernst feared that her lawsuit was over and would be "destroyed." She heard the news while preparing her presentation for the United Nations Commission on Sustainable Development. UNANIMA International, a Catholic group with 78,000 members representing seven female religious orders, had arranged the invitation for Ernst to speak there. The group, which promoted the "feminine life principle of healing, caring and nurturing," thought the UN needed to hear a woman's firsthand account of what it was like to deal with the hydraulic fracturing industry. One of UNANIMA's members had Googled "Canada's Erin Brockovich," and that's how the religious group discovered Jessica Ernst.

As the Canadian election results rolled in, Ernst cried in a tiny room at Leo House, a Catholic guesthouse in New York's Chelsea neighborhood. She practiced her presentation once more and then lay down, but she didn't sleep—she never sleeps before a presentation. In the morning, she vowed to keep going, no matter what Harper did to her or the case.

In her short talk at the United Nations, Ernst argued that fracking could never be sustainable because "it poisons water and divides communities, and captures our energy regulators and elected officials." Prior to the event, a Canadian reporter had asked her why she was going to New York—"to give Alberta a black eye?" Ernst got pissed off: "I'm not giving anyone a black eye," she snapped. "Alberta is giving itself a black eye."

JUST AFTER ERNST returned to Rosebud, Duke University released an explosive study on private water wells near fracked drilling operations in northeastern Pennsylvania. Researchers found that groundwater wells a mile away from the fracked gas wells contained seventeen times more methane than water wells farther away. Leaky wellbores, industry-made fractures, or the expansion of existing fractures probably accounted for the gas straying into water wells, the researchers concluded. *National Geographic* illustrated the story with a photo of Jessica Ernst lighting up a five-gallon water jug full of her methane- rich tap water in the safety of her barn.

By now, Ernst was following the protracted legal journeys of other North American groundwater contamination cases. Since the shale gas boom had begun in 2005, dozens of cases had popped up, in Pennsylvania, Texas, Colorado, Arkansas, and Louisiana. In 2011, the San Francisco–based Sedgwick law firm reported that hydraulic fracturing litigation had become a major legal trend. In Texas, multiple lawsuits had arisen from fracking operations in the Barnett Shale alone, "with plaintiffs complaining of flammable water, violations of the Safe Drinking Water Act and discolored, sediment-filled water." Not all the cases focused on groundwater contamination, either. In Arkansas and Texas, residents sued the operators of compressor stations for "emitting large quantities of noxious gases along with producing loud noise damaging the plaintiff's hearing." In Arkansas, citizens also sued major energy companies for causing "an unprecedented increase in earthquakes" with their disposal wells. The quakes left "property damage, real estate devaluation and emotional distress." The Sedgwick report added that the litigation was "taking on an international character, with a recent case filed in Canada."

Ernst noticed a worrisome development in the lawsuits, something Texas blogger Sharon Wilson later described as "the cycle of fracking denial." Regulators began by claiming there was no proof of groundwater contamination. When landowners provided proof of methane or hydrocarbon contamination, industry attempted to bury it by offering landowners cash in return for signing confidentiality agreements. Landowner Grace Mitchell, for example, had sued Encana in 2010 in Johnson County, Texas. After Encana fracked shales near her property, Mitchell could "no longer use the water from her own well for consumption, bathing, or washing clothes because in approximately May 2010, the well water started to feel slick to the touch and give off an oily, gasoline-like odor." Mitchell settled out of court and went silent. Even court discovery materials in her case were subject to "a protective order." Gag orders erased history, Ernst realized, and allowed regulators to claim there had been no proof of contamination in the first place. To her way of thinking, the courts were participating in "criminal activity" by allowing the gag orders. She had compassion for families who signed to protect the health of their children but only contempt for the authorities that willfully covered up industry's dangerous methane liabilities.

In October Ernst returned to New York at UNAMIMA's invitation. This time the group presented Ernst with the International Woman of Courage Award for "her struggle to overcome the injustices brought about by the predatory search for profits at whatever cost to present and future generations." During the ceremony, Ernst cried and was too nervous to eat. The Alberta Government never recognized the achievement.

In November 2011, the ERCB abandoned its in-house legal team and hired the high-profile Calgary law firm Jensen Shawa Solomon Duguid Hawkes (JSS) to direct its defense against Ernst's lawsuit. The "civil litigation boutique" boasted impeccable ties to both the Conservative Party of Canada and the Alberta government. One of the firm's principals, Robert Hawkes, was the former husband of then Alberta premier Alison Redford, and he remained one of Redford's trusted

political advisers and campaigners. While serving as Alberta's justice minister in 2010, Redford had personally chosen
her ex-husband's law firm to handle a $10 billion tobacco law-
suit on behalf of the government. (An ethics investigation
later cleared Redford on a technicality.) JSS handled business
for several energy firms, including a former Encana entity:
Cenovus Energy. The firm also represented the Office of the
Information and Privacy Commissioner, which Ernst had now
been battling for four years. Most critically for Ernst's lawsuit,
JSS did work for the Harper government.

A month after Alison Redford became premier, JSS senior
partner Glenn Solomon got the job of defending the ERCB.
Solomon, an energy litigation star in Alberta, had known Redford for twenty years. He not only donated regularly to the
Conservatives but had served as a director of several federal
Conservative Party riding associations. JSS celebrated Solomon's "political involvement" on its website, alongside many
glowing peer reviews of his legal performance. To Ernst, Solomon's involvement in her case was a "fitting" reminder of the
threat her lawsuit posed to a brute-force technology and its
advocates.

Solomon took a hardball approach to defending his client.
His first legal brief, some seventy-three pages long, contended
that the ERCB owed no legal duty of care to protect a citizen's
groundwater—but it did have a public duty to implement
legislation for the oil and gas industry. Although the Charter of Rights granted citizens the freedom to thought, belief,
and expression, Solomon's brief argued, it did not "guarantee
the right to be listened to." Furthermore, the Charter "does
not guarantee an audience." As a consequence, the brief
said, the regulator had been within its rights to suspend

communication with Ernst. Finally, said Solomon, even if the ERCB was negligent, Section 43 of the Energy Resources Conservation Act—the immunity clause—exempted the regulator from any liability. "It is obvious that the claims against the Defendant are without merit," concluded Solomon.

Klippenstein had expected Solomon's line of reasoning, but he found the defender's arguments extreme. "I think most Albertans would not be comfortable with a regulator that says it is basically immune from legal accountability in a democracy no matter how incompetent and negligent they are," he told a reporter. "That's a very unusual position for a regulator." Klippenstein pointed out that a coal-mine regulator owes a duty of care to miners, to ensure that their workplace is safe, and that municipalities owe a duty of care to their residents, to ensure that building codes are enforced. Why should an oil and gas regulator not be held accountable for "negligent failure to comply with established government policy?" he asked.

In contrast to Solomon's lengthy, convoluted briefs, Encana's lawyers filed a simple thirty-three-page brief. To Ernst and her lawyer's astonishment, Encana's lawyers didn't propose to dismiss the entire lawsuit or press for summary judgment—the usual course chosen by powerful corporations. (The notoriously litigious company had once tried unsuccessfully to sue a Colorado bluegrass group for a song that asked, "How did they poison my water and hay, by drilling for gas in the ground?") But now Encana merely wanted to strike 130 paragraphs from Ernst's claim on the grounds they were frivolous or too detailed.

To guide its defense, Encana hired Maureen Killoran, a managing partner in Osler, Hoskin & Harcourt, one of Canada's biggest firms. Killoran, the mother of four children, told the *Calgary Herald* that "when you do energy law, as I do, or corporate litigation, you're not dealing with life and death situations and people whose lives have been turned upside down, plaintiffs who are weeping. It's just about money."

The lawyers representing Alberta Environment initially registered no objection to the lawsuit. But they requested that more than a hundred paragraphs be struck from Ernst's claim because it read as a "critique of the oil and gas industry." Their brief argued that any "allegations regarding how the oil and gas industry as a whole operates or how the oil and gas industry may affect other individuals are irrelevant, improper and should be struck."

On April 16, 2012, ten days prior to Ernst's first hearing in the Drumheller Court of Queen's Bench, the province's OIP Commissioner unexpectedly opened a door for Ernst. In a scathing ruling, Adjudicator Teresa Cunningham ordered the Alberta Research Council (by then renamed Alberta Innovates–Technology Futures) to release six thousand pages of documents it had censored or withheld relating to investigations of the causes of methane contamination in water wells. Cunningham's sixty-five-page ruling found that ARC had wrongly charged Jessica Ernst $4,125 for a request that should have cost $295 and that the outrageous fees were intended to punish a citizen for requesting public data. Furthermore, it described ARC's four-year obstruction of information as having "the effect of undermining a central purpose of the FOIP Act: the right of timely access to records in the custody or control of a Public Body."

Ernst still considers the OIP ruling her most important legal victory. That same month, the Alberta government again ranked last in a transparency study done by the Halifax-based

Centre for Law and Democracy. The center noted that, on an international comparison, Alberta's administration of right-to-information laws placed the province somewhere behind Angola, Colombia, and Niger. (Under Stephen Harper's leadership, Canada by now placed behind Mongolia on information transparency.)

Ernst's lawyers were dumbfounded by the contents of the documents ARC had withheld for four years. One damning find was a 2007 "Coalbed Methane Complaint Response Review" that had never been released to the public. The review—from which ARC still withheld two chapters—recommended that citizens unsatisfied with the province's groundwater review process be able to pursue "civil action." A batch of government emails also proved that ARC's summary and the five individual reports on groundwater contamination were anything but "independent." Just as Ernst suspected, the ARC report had been heavily and secretly edited by the agency that contracted it: Alberta Environment. Two years later, these documents would save Ernst's lawsuit from oblivion.

TWELVE

The Road
of the Dishes

IN THE EARLY days of 2012, North America's fracking industry was eying Europe with plans for expansion. Since 2006, the controversial technology had doubled the number of natural gas wells in the United States, from 342,000 to more than half a million. Companies such as Chesapeake Energy and Encana had amassed leases greater than the size of West Virginia to crack rock. About 20 percent of Colorado had fallen under the control of frackers, mostly the slick-water variety. More than 15 million North Americans lived within a mile of a well site fracked for gas or oil. Jubilant analysts predicted that "the biggest innovation in energy so far in this century" might even restore the United States as the globe's number one energy producer.

The initial numbers seemed impressive. In just a decade, the market share of unconventional gas had gone from zero to 37 percent. In fact, the fracking of low-quality rocks in North America had temporarily created a glut, lowering natural gas prices from $14 to $2 a unit. The frackers had also poured 4 million barrels of oil from the Bakken Formation and the Eagle Ford Shale into North American refineries. The horsepower needed to crack these shale basins exceeded the energy output of eleven medium-sized nuclear reactors and often cost more than the product itself. But no matter. Wasn't it time to take what engineering textbooks described as a "technology for all time and all places" to the rest of the world? The shales of Ireland, Poland, China, Argentina, South Africa, and Ukraine beckoned with the promise of higher gas prices and fewer regulations.

Industry promoters in Europe promised lots of money from fracking, but like their North American counterparts, they left out critical details. For starters, they failed to mention rapid well depletion rates over a period of three years—80 percent—that put the industry on a costly and endless drilling treadmill. The spin doctors also glossed over dismal recovery rates of less than 10 percent. Cumulative effects were rarely mentioned. Air pollution from fracking well sites in Wyoming, for example, choked rural populations with ozone pollution worse than that found in Los Angeles. Truck traffic in Texas had destroyed a billion dollars' worth of public roads in the Barnett Shale, while local counties earned a third of that in industry taxes. And in British Columbia, scene of some of the world's largest frack jobs, industry had set off hundreds of small earthquakes in the Horn River Formation.

The Irish got the full sales pitch. Tamboran Resources, an Australian firm, advertised fracking as a "100 per cent safe" activity with absolutely no risk. The technology created only "opportunities." By cracking rock under a million hectares in the Lough Allen shale basin along the border of the republic and Northern Ireland, both governments could get lucky and fill their depleted coffers, promised CEO Richard Moorman, a Calgary-based engineer. The basin might also provide the island with twelve years' worth of methane. The technology, explained Moorman, "works like a crow bar. It just cracks the rocks open and it is not a very fancy process when it comes down to it."

235

But Ireland's mothers and fathers didn't believe "Richard Moorman's roadshow," as Meg Rybicki dubbed it. Having survived a financial crash during which every authority and professional lied to the public, the Irish suspected the frackers were just another breed of foreign pirates. So, too, did Irish-born doctor John O'Connor. The family physician worked in northern Alberta, and he had drawn attention to rising rates of rare cancers downstream from the huge tar sands project. O'Connor had witnessed how an oil boom could undo a place. When friends in Ireland asked him about the fracking experience in Canada, O'Connor told them about Jessica Ernst.

In January 2012, O'Connor asked Ernst if she would tell her fracking story on the Emerald Isle. He offered to pay for the flight. At first Ernst said no—she didn't want to leave home or deal with the crowds. But O'Connor persisted, arguing that the Irish needed to hear a firsthand experience. In the end Ernst agreed, believing that life often boils down to how much you can give. On both sides of the Irish border, local groups arranged information sessions.

The unexpected invitation caught Ernst at an auspicious moment: she was just finishing the research on a lengthy paper about gas migration. Since 2005, she had uncovered a largely hidden literature on leaky wells and straying gases. More than a thousand articles outlined a serious industry liability that fracking worsened: just about everything industry injected into the ground had a hard time staying put, whether it was frack fluid, stored gas, or dirty saltwater. The science also portrayed gas migration from wells as a chronic problem, with predicted rates of occurrence. As a consequence, aging leaking wells almost guaranteed that, as one 2000 paper put it, "the concentration of gases in the shallow aquifers will increase with time."

Not all methane escaped up or along badly sealed wellbores, either; it also migrated horizontally. "There are many cases of leakage where the gas simply enters the water aquifer and may never bubble around the casing," explained one study. All in all, Ernst learned that gas migration was to the oil and gas industry what superbug infections were to hospitals—a nasty nemesis that professionals largely disavowed and regulators denied. A 2003 *Oilfield Review* article summed up the problem with classic understatement: "Successful handling of gas migration is an evolving science." Ireland seemed like the right place to share these and other damnable findings.

When Richard Moorman heard about Ernst's upcoming visit, he quickly tried to discredit the scientist. (Members of Love Leitrim said when they met Moorman and Tamboran officials, "their anger was palpable" whenever Ernst's name came up.) The petroleum engineer told Irish national television that there was only "one woman" in Alberta who had any problems with fracking. Representatives of the Alberta government later called and told members of Northern Ireland's parliament the same thing. Nevertheless, the Irish came out in droves to hear Ernst.

Her first hour-and-a-half-long talk, "Life Inside a Frac Experiment," took place in the village of Belcoo in County

Fermanagh, scene of many bloody troubles. To save time, Ernst thought she'd zip through part of her story, but the audience wouldn't have it. Two rosy-cheeked farm women protested loudly. They had seen an Alberta talk by Ernst on YouTube, and they demanded the long version. "Do you want me to add the whole police bit?" asked Ernst. The audience yelled, "Please!" An intense silence descended on the room as Ernst recounted the intimidating visit from the RCMP. The crowd burst into delighted laughter when Ernst got to the point where the constable blurted, "You can't take our photo, we're undercover." (An Irish Conservative MP later protested that such overt police intimidation could never happen in a gentle place like Canada.)

Several days later, more than 350 people filled the Rainbow Ballroom of Romance in Glenfarne, a majestic place scheduled for industrial fracking. As a reporter recorded, "Local men in the doorway strained to hear the words of the speaker as if their lives depended on it."

Ernst opened her talk by admitting that it was really hard to resist and to speak out. But on the issue of fracking, she said, people had to let their fears go. "I'm here to tell you tonight that laws and regulations do not protect us from the new brute force of hydraulic fracturing or the new 'super fracking' experiments."

Ernst peppered her story with nuggets from her research on gas migration. Even before the frackers came along, she said, the industry had a costly liability on its hands: leaking wells. The problem got worse as the cement aged and as industry punctured more pathways into the earth, providing more opportunities for gas to migrate. A 1986 Saskatchewan study that surveyed 939 water wells and springs found that

"methane concentrations were highest where petroleum industry drill hole density increased," she told her audience. And
that was before the advent of massive high-volume fracking.

238   Ernst traced the history of fractures going out of zone
into freshwater aquifers and warned the assembled group
that fracks were unpredictable things: they didn't stay in the
target zone, and they followed the path of least resistance.
No amount of industry denial could change that fact. Next
she showed her attentive audience the cover of the 1987 EPA
report to Congress that had documented how "residual fracturing fluid migrated into a water well" in West Virginia in 1982.
As the EPA later admitted, and the New York Times would report,
hundreds of other cases had been hidden by confidentiality
agreements or gag orders. No one had the right to cover up
contamination of lakes and rivers, said Ernst calmly, "because
we share our water." There was a groan of recognition.

Ever the scientist, Ernst methodically piled on more evidence, including a 1989 paper that found hydraulic fractures
in several energy wells propagating "into an underlying water
zone" in Manitoba's oil fields. Next came a 1993 Husky study
that described gas migration as "a big problem" that was
"expensive to fix and difficult to stop." In fact, half of Husky's
20,000 deviated wells leaked. Ernst included, too, the U.S. Geological Survey research on leaky gas wells in the San Juan Basin.
"They didn't find that nature was causing the upward migration
of gas," replied Ernst. Instead, the study concluded that "gas
wells are more important than nature for the upward moving
of gas." Ernst marveled at the audience's concentration and
then teased them: "Learning about fracking on a Friday night,
I can't believe you're Irish."

After that moment of comic relief, she continued. When
the Canadian Association of Petroleum Producers investigated

gas migration in 1995, she told her listeners, they titled their report "The Migration of Methane into Groundwater from Leaking Energy Wells." "They didn't say 'migration of methane into groundwater from nature,'" Ernst pointed out. And when the 2002 study by the Canadian Council of Ministers of the Environment identified the hazards posed by unconventional drilling to groundwater, "they didn't say nature is going to contaminate your water," said Ernst. The study warned that industry would.

To illustrate for her audience how wayward fracks could travel, Ernst used an incident in Hutchinson, Kansas, as an example. In 2002, methane from the underground Yaggy Storage Field gas facility escaped via a leaky injection well into natural fractures. The gas traveled nearly seven miles underground and came to surface in old salt wells in the middle of town, where it exploded, blowing apart two buildings. Several thirty-foot geysers of salt and gas later erupted two miles away. A cloud of methane that ignited near one of the geysers killed two people. Fracking in Glenfarme, Ernst noted, had the potential to propagate into rivers and lakes miles away.

"Will I sign a nondisclosure agreement and allow the Rosebud water well records to be sealed?" asked Ernst near the end of her talk. No, she said: "I will not sign one." No matter how you spell it, she warned, "fracking brings a war that knows no borders." She encouraged her audience to argue things out and then to "prove to the world that there is a country that this industry cannot divide... When you are fracked, there is no aftercare." A host of questions followed.

By the close of Ernst's tour, the Irish were calling her the "Joan of Arc of Alberta" and a "Rachel Carson of the Environment." Ernst spoke directly to thousands of people, and Ireland's national network interviewed the oil-patch scientist three times on television. Community groups booked her for a return visit the following year. "Forewarned is forearmed," wrote a grateful physician colleague of O'Connor's. To this day, O'Connor swears it was the best investment he ever made. As for Ernst, she arrived back in Alberta exhausted and was sick for three weeks.

Several months later, Tamboran Resources changed its "100 per cent safe" position on fracking. In a June presentation to the Northern Ireland Assembly, CEO Richard Moorman admitted that some projects in the United States had punctured landscapes with as many as one hundred wells in the space of a mile and concluded that yes, "you will have a surface impact." Hydrocarbons, lamented the engineer, "are in their own form of twilight." Shale might be the last rock mined, and for good reason, he said: "You would have an easier time getting gas out of your sidewalk than through what we try to do by cracking shale." Regulators, Moorman agreed, needed to "enforce better than they have been." Not long afterward, Moorman became the "former" CEO of Tamboran Resources.

As news of Ernst's lawsuit spread, invitations to speak flooded her inbox. In May 2012 she flew to Michigan, where she talked to hundreds of landowners about the truth and consequences of fracking. Encana had attacked the Antrim Shale there, and problems galore had arisen. An industry representative stormed out of one of her talks, shouting that it was 80 percent embellishment. "That's funny," responded Ernst. "Eighty percent of my talk is based on industry and government data, and only 20 percent is personal." While in Michigan, she also explained how industry undermined communities by promising them money before they fracked them. "No healthy community will allow hydraulic fracturing," said Ernst, "so they [the industry] have to make the community sick. And they do so by feeding the dark side [of] human nature, which is greed, sloth, selfishness. They feed the ego, they promise a little bit... and then whammo, the community is divided. The people with concerns are then abused by the people who want more money, and Encana doesn't even have to do the dirty work."

241

After Michigan, Ernst traveled to upper New York State, which had declared a moratorium on fracking in 2011. Audiences there couldn't believe that she wouldn't sign a gag order. During her talks, she teamed up with Cornell University fracking expert Anthony Ingraffea, whom she had met a year earlier at a talk in Halifax, Nova Scotia. She and Ingraffea had corresponded since then on the technical nitty-gritty of fracking. The two scientists made an odd pair. Ernst was living a life that had been fractured, and Ingraffea, a world authority on the mechanics of fractures, knew that her story reflected the hard reality of the science.

Like Ernst, Ingraffea was a consummate industry insider. For more than thirty years the diminutive scientist and engineer had studied the nonlinear propagation of fractures in rock, concrete, aluminum airplanes, and dams. As director of the Cornell Fracture Group, Ingraffea had also supervised $35 million worth of studies on fracture behavior for the likes of Exxon, Schlumberger, and Boeing. He still served as editor of *Engineering Fracture Mechanics*. Articles published in the journal routinely admitted that "the propagation of fluid-driven fractures in a porous medium is an important problem in rock mechanics" and a problem all about "moving boundaries." Ingraffea understood perhaps better than anyone else how intense periods of high stress invariably led to unpredictable fractures.

Since 2009, Ingraffea had also become a major critic of the fracking industry. He had concluded that policy on the technology clearly outpaced the science and that he had a responsibility to challenge industry's lies. In addition, he had intimately studied the technology's fatal flaw: the inability to contain an induced fracture. Early in his career, Ingraffea had been part of a Lawrence Livermore National Laboratory team that tried, in 1983, to develop computer simulation models to show how fracture fluids might travel through naturally fractured Devonian shale rocks. To get the shale to let go of the methane, industry had to connect to existing natural joints, cleats, and fractures with a multitude of man-made fractures over a vast area. The team also tried 2D and 3D computer models to illustrate the complexity of the connections, but to no avail. "The fracture wanted to go wherever it wanted to go," recalls Ingraffea. "You couldn't predict it." Thirty-two years later, Ingraffea still gets calls from petroleum executives asking him for help in producing an accurate predictive model for fracking. "That tells you everything you need to know," he says. "They are just using brute force."

To Ingraffea, the fracking of underground rock is, like earthquakes or climate change, a perfect example of nonlinear chaos. Even in the best-case scenario—with a vertical fracture in tight sandstone "where you create butterfly-like networks of fracks on either side of the wellbore," you can't know what's going to happen, explains the engineer. Natural fractures in shale rocks just compound the chaos. A two-mile-long lateral from a horizontal well might connect to millions of natural fractures. "Which ones are going to be stimulated by the process? It's a classic problem of mathematical chaos." Coals offer engineers more challenges. "They are more fractured than shale, rest in potable water and lie at shallow depths," Ingraffea points out.

Fractures behave so randomly that even teaching PhD students how a fracture spreads and grows remains a headache, says Ingraffea. "You have to imagine yourself thinking in 3D," he explains, just to do the process justice. "Imagine a cube made of bricks. Now imagine all the joints are really natural fractures. Next imagine a fluid hitting the cube at high speed in a short burst. Try to visualize what happens. Where does the fluid go? It goes where the hell it wants to go, and that's what I visualize... Even in the best of cases, and shale is not the best, it is impossible to predict the results of a hydraulic fracturing stimulation."

While in New York State, Ernst got a chance to interview Ingraffea about fracking, regulators, and the perils of gas migration. On May 23, the two sat on a bench in a garden at Cornell University and chatted away like old colleagues. Ernst played the role of an inquisitive and informed reporter.

"So, Tony, tell me," she began. "What do you think about blaming nature all the time for water contamination?"

Well, said Ingraffea, it was an absurd habit, and a fine example of what Samuel Taylor Coleridge once called the "willing suspension of disbelief." "Somebody could be cruising along on their own perfectly clean well water for generations, and suddenly gas operators come to town and drills scores of wells within few kilometers of somebody's water well, and it goes bad. What a tremendous coincidence that it was naturally generated biogenic methane that happened to find itself, at the very same time, into a water well that it has avoided for hundreds of years!"

Why was industry now blaming the contamination of water from deep wellbore gas on nature, too? Ernst asked next.

"Well," answered Ingraffea, "they can because they have to. And you know why they have to."

"Because of billions of dollars of liability," interjected Ernst. "And explosions, death, and multi-million-dollar explosions."

"And don't forget that they have hidden all of this from the rest of the public," added Ingraffea, "by way of doing what you say you will never do, which is to sign a nondisclosure agreement."

As their interview continued, Ingraffea explained how methane migration, loss of wellbore integrity, and contamination of underground sources of drinking water had dogged the industry from its infancy. "And now we are facing what our world leaders are calling the Golden Age of Gas, in which we are going to see, worldwide, millions of new wells drilled, using a technology that does not decrease the likelihood of all of these phenomena but substantially increases their likelihood. And yet they still want us to believe, whenever someone complains about contamination of their groundwater, that it was just a fantastic, natural coincidence."

What about the cumulative effects of repeated perforations and fracking on the cementing and sealing of a wellbore? queried Ernst.

That was a huge problem too, said Ingraffea. "The new kind of high-volume fracking, which hopefully we won't see in New York, exacerbates and makes worse a bad situation. It used to be with the old conventional wells there was a well and you'd have to go maybe a kilometer away to the next well. And maybe you were lucky enough not to see the chemical constituents that went down one well affecting the integrity of the cement on that well. But now we are seeing many wells drilled from one pad and many wells restimulated adjacent to existing wells, and both those processes, pardon the pun, put more pressure on the cement. The strength of the cement, the toughness of the cement, and the durability of the cement are all very important qualities you want, and are all being challenged by the mechanical process of joining one well very close to another well and all the ground motions that occur." Thanks to hydraulic fracturing, added Ingraffea, the industry is speeding up a cement degradation process "that used to take decades." Now, he said, it takes only years.

The two scientists then joked about which government regulator had established the most effective practices for industry to follow. Alberta? Pennsylvania? Australia? Best practices, Ingraffea said more seriously, were just an industry ploy "to minimize liability for the things that we know will go wrong at an increasing rate."

Given the failure of regulators and self-regulation by industry, Ernst asked Ingraffea next, how can civil society restore sanity to the process?

"I'm a realist, a pragmatist, an engineer, and a scientist," replied Ingraffea. "Where unconventional gas is not happening, stop it from happening. That's only realistic, given the track record we've seen so far from an industry that said, 'Don't worry, we've been doing this for sixty years.' Now they have spent the last four years trying to develop best practices? I would say you can't trust them. Where [unconventional gas extraction is] already happening, slow it down as best as you can. Then demand the best regulations, rigorous monitoring, and up-to-date technological practices such as better cementing."

Ernst's last question was a stunner that captured the full gravity of the topic: "How many holes can industry put in the earth? Is there a limit?"

Ingraffea called the question one of those "absolute cardinal problems" in science. Unconventional resources demanded

more holes, and that would have profound geological consequences. "If you just drill one hole and one lateral and frack it,
and claim that you now understand all the geological implications, all the probabilities, and all the risks of something bad
happening, you are fooling yourself. You are looking at a scale
which really isn't accurate. One well, one lateral, one frack can't
answer the question you asked."

What happens when an industry plants 100,000 wells in a
region, a country, or a state and then stresses every wellbore
with a high-volume frack job? asked Ingraffea. "Haven't we
really damaged Mother Nature [such] that she might not be
able to repair herself or to such a significant degree that there
is no hope that even geo-engineering can right the wrongs? I
think we are on the path to doing that experiment in real time
with us in the test tube."

Later that year, a group of geological engineers with Con-
ocoPhillips unwittingly supported Ingraffea's conclusions by
again highlighting the nonlinear behavior of hydraulic fractures. In the *American Association of Petroleum Geologists Bulletin*,
the engineers complained that computer models could still
not explain why fractures made so many complex shapes.
They described the cracking of underground rocks as "a complex process" that "may enhance, redirect or suppress hydraulic
fracture propagation."

ONCE SHE POSTED it to her website, Ernst's gas migration
paper was viewed by hundreds of scientists and government
researchers around the world. Copies went to the New York
Department of Environmental Conservation and New York
State governor Mario Cuomo. The government of Quebec,
which would later limit fracking in the province, requested a

copy too. In Newfoundland, concerned citizens placed a copy of Ernst's report in every library. The government of New South Wales asked Ernst for a copy, but then omitted the subject of gas migration in its final report on fracking, naming the Alberta government as a promising pioneer in regulation.

In March 2013, Ernst returned to Ireland, where farming communities were now actively challenging politicians seduced by the frackers. One night during her week-long tour, which also included two talks in England, Ernst was hosted by a family who lived in the Burren, by the sea. Great limestone formations haunt the ghostly landscape. A member of Oliver Cromwell's army had described the place in 1651 as not having enough water "to drown a man, wood enough to hang one, nor earth enough to bury them." Now industry wanted to frack the storied landscape where tufts of grass between the limestone still fattened cattle.

In the Burren, Ernst stayed with Martina O'Dea and members of her cattle-farming family. In a four-hundred-year-old farmhouse, Ernst sat with O'Dea's relatives by the fire. The women chatted all night long and talked of the dead that walked the ancient land. O'Dea felt the dead were rising to stop the crazy business of fracking.

At one point, while O'Dea was driving Ernst back to the farmhouse, the subject of scarcity came up. Ernst feared for the future, she told O'Dea, and she wondered if she would have the resources to pay her legal fees to the end. O'Dea smiled and informed Ernst that they were on the Road of the Dishes. "There is a myth about this place," O'Dea told her passenger. "Would you like to hear it?"

Hundreds of years ago, said O'Dea, St. Colman had lived in a nearby cave with his assistant. After fasting for Lent, both men were starving. Nearby, at his castle, King Guaire had prepared a big Easter feast. When the saint exercised his faith, saying, "Dishes rise. Come to me," the dishes rose off the king's table and flew down the road through a mountain gorge to the saint's cave. When the angry king ordered his soldiers to pursue the dishes, their horses' hooves got stuck in the crag, and so the saint and his assistant sat down to a fine Easter meal. Now, whenever Ernst gets scared about the cost and duration of the her lawsuit, she remembers the Road of the Dishes.

THIRTEEN

"No Duty of Care"

O N APRIL 26, 2012, a cold spring day, Jessica Ernst awoke
with a lick on her nose from Magic. "It made me laugh,"
recalls Ernst. "We are ready, I thought."

Five years after filing her lawsuit, Ernst finally entered a
Drumheller courtroom. Landowners, including the Campbells
and Fiona Lauridsen, turned out to more than half-fill the concrete room. As one central Alberta farmer bluntly put it, "This
may be the case that determines whether Alberta becomes
a dried-up, toxic, smoking black hole in thirty years or not."
When Justice Barbara Veldhuis entered the austere courtroom,
Ernst gasped in astonishment. The tall, striking woman with
long, straight hair looked like Ernst's doppelganger.

Lawyers for the defendants, assembled on the right side of
the courtroom, had come ready for a prolonged legal face-off.
Glenn Solomon stated he was prepared to argue for the dis-
missal of the entire case against his client, the ERCB: "The ERCB
does not engage in, and is not alleged to engage in, drilling for
shallow coalbed methane." Lawyers for Encana and Alberta

Environment wanted to censor half of Ernst's statement of claim due to its "inflammatory" and "embarrassing" contents. ("Yeah," Klippenstein said in an aside to Ernst. "Embarrassing for whom?")

But Justice Veldhuis had other plans. Given the complexity of the claim, the judge offered to serve as the lawsuit's case manager. All the parties agreed. Then Veldhuis said she might do something different, and she did: she requested a shorter statement of claim. In response, Ernst's lawyers asked for a recess to consult with their client in a back room. Klippenstein and Wanless didn't think a shorter claim was a smart idea, and explained it as a stalling tactic designed to punish the plaintiff financially. Ernst countered that agreeing to a shorter claim might be useful: her adversaries wouldn't expect it. Back in the courtroom, while consenting to Veldhuis's request, Klippenstein predicted that at some point someone would complain that there was not enough information in the shorter claim. (That someone would turn out to be the Alberta Court of Appeal, two years later.)

Once Klippensein had agreed to the judge's request on behalf of his client, Solomon asked Justice Veldhuis to remove the original claim from both the court system and the public record. Veldhuis, however, denied what Ernst called "the ERCB's hideous request." The original statement of claim remained on her website, ernstversusencana.ca, and is well read to this day.

In early June, Ernst's lawyers submitted what the gatekeepers wanted: a pithy twenty-six-page statement of claim. But the gatekeepers were still not satisfied. During a case-management conference call six months later, Solomon argued that it wasn't convenient to have the case heard in Drumheller. "None of the counsel in this case are in Drumheller," reasoned Solomon. "The judge is not in Drumheller. The only person near Drumheller is Ms. Ernst. We are asking her to travel to Calgary if she wishes to attend in person. That's about 100 kilometres from Rosebud, where she lives, which is between Calgary and Drumheller, and I submit that having one person travel 100 kilometres, rather than having seven or eight or ten people travel 140 kilometres, is in the interests of efficiency." Klippenstein protested that his client had the right by law to have her case heard in the court closest to her and to where the harm occurred. In fact, court rules said as much. But a lawyer from the Alberta government seconded Solomon's proposal, and the Encana lawyers advised that they did not want to drive to Drumheller, either.

When the Chief Justice of Alberta's Court of Queen's Bench approved the unorthodox decision to move the upcoming court hearing to Calgary, Ernst felt "sicker than sick." She considered the move not only unfair but disrespectful to rural Albertans. It also contravened basic court rules. She asked Klippenstein to appeal the decision, but he reasoned it might look bad or turn the judge against her. "Pick your battles," said Klippenstein. So Ernst began to devise her own plan.

In preparation for the second procedural hearing, Solomon honed his arguments in defense of his client, the ERCB, in two more legal briefs. He portrayed Ernst's claim as frivolous and argued that the board "owed no duty of care" to the landowner. Moreover, Section 43 of the Energy Resources Conservation Act gave the board blanket immunity. Solomon's brief also characterized Ernst as an "eco-terrorist." He claimed that the board had stopped talking with the oil-patch consultant solely because of comments about "the Wiebo Way." Faced with "a real threat of violence," Solomon's brief said, the board could do only one thing: end all contact with the errant citizen. According to Solomon's selective version of events, the board had banned communication with Ernst "in order to protect its staff, the Alberta public, and the Alberta oil and gas industry from further acts of eco-terrorism." By threatening violence against authorities, added Solomon, Ernst had invalidated her Charter claim.

In their own legal briefs, Ernst's lawyers replied to the slander: "If the ERCB wishes to advance its patently absurd and irresponsible theory that Ms. Ernst's offhand reference to Wiebo Ludwig was somehow a 'threat of violence,' and that an appropriate response to 'protect against further acts of eco-terrorism' was to cease communication with the Plaintiff, it must do so by forwarding cogent evidence. The ERCB has not, and frankly cannot, put forward such evidence."

While the lawyers were exchanging their missives in the court system, an ERCB researcher reported at the annual Unconventional Resource Conference–Canada that industry high-volume horizontal fracks had "communicated" with neighboring oil and gas wells nearly 40 times over the previous three years. The fracks had traveled an average of 1,300 feet, and in some cases as far as 7,000 feet. As a consequence, frack fluids had exploded up nearby oil and gas wells. The ERCB study, which offered no insight into how industry-made fractures might impact groundwater, abandoned wells, or vertical wells, was never posted on the board's website.

Ernst's second court hearing took place in downtown Calgary on January 18, 2013. Despite the change of location, as many as eighty Alberta landowners and citizens packed the courtroom, an unprecedented number for an infant lawsuit.

Justice Veldhuis told the overflowing courtroom that she wouldn't be making a decision that day to either drop the ERCB from the case or to strike out parts of the claim. The lawsuit was just too complex, she explained, and it came with a   253
daunting volume of reading material.

As the Calgary hearing proceeded, there was no sight of Ernst in the room. That morning, while an Encana rig completed another gas well in the Rosebud Valley, Ernst had driven to the Drumheller courthouse. Court rules entitled her to a hearing where the harm took place, she said, and she wanted the law upheld. Together with a witness, Shirley Bray, she sat where the polite and supportive clerks instructed her to sit: in the waiting room. She sat there all day. Bray later called it the strangest day of her life.

At one point, an RCMP constable introduced himself. "Aren't you Jessica Ernst?" he asked.

Ernst said yes and shook his hand.

"My parents and I, we love you!" said the constable. "Aren't you supposed to be in Calgary right now?"

"Nope," said Ernst. "I am supposed to be right here." She explained why.

The constable informed Ernst that the water well at his parents' place had gone bad in 2002 after nearby fracking. When the couple asked the regulators to do their jobs, they got nowhere. Ernst couldn't believe how fractured the world had become.

Back in Calgary, Klippenstein explained Ernst's curious absence to the court. He read a note from the missing plaintiff into the record: "I... feel strongly that my lawsuit should be heard in the judicial district of Drumheller... This is where I live; this is where my water is; this is where coalbed methane wells were drilled; and this is where my water is contaminated. It seems to me that it is important to rural Albertans that disputes and harms that occur in our communities are also judged in our communities. Today's application is not a minor or merely procedural step—it will determine the core issue of whether a landowner can sue the Energy Regulator for failing to protect rural Albertans from the harmful effects of the oil and gas industry."

Throughout the all-day hearing, Encana's lawyers said virtually nothing. They hadn't even filed a defense yet. They appeared content to let Glenn Solomon do the heavy lifting as he hammered away on his no-duty-of-care arguments. "Ms. Ernst's interests cannot prevail over the public duty," declared Solomon. And even if they did, the ERCB's immunity clause made such matters irrelevant. Solomon suggested that if Ernst had focused her claim on Encana, instead of trying to sue the regulators, "she'd be through trial by now." Encana, after all, was the regulated party; if they had acted contrary to the law, then Ms. Ernst should get relief from them, Solomon argued. "But she doesn't get relief from the ERCB on the basis that they didn't adequately or at all regulate Encana because the ERCB is not providing a direct service to Ms. Ernst in priority to its public duties. And it cannot." Without statutory immunity, no energy regulator could discharge its duties effectively, expounded Solomon. Allowing regulators to be sued for negligence would be like letting people sue judges because they didn't like a particular court ruling. "If Ms. Ernst doesn't like your decision, she doesn't get to sue you," he told Justice Veldhuis. "She gets to appeal you."

Murray Klippenstein reminded the judge that there were two issues before the court: a Charter case relating to freedom of expression and a negligence case based on what the ERCB and Alberta Environment had failed to do. The judge might find the board immune from one challenge, but not both, he said. To emphasize the importance of freedom of speech in the case, Klippenstein retold the banning story. "What Ms. Ernst is saying is that the Board has to at least receive her expression as a basic minimal threshold level of her right of expression. They cannot slam the door in her face and walk away because they feel like it. And that's what happened." In democracies, Klippenstein argued, government agencies can't pick and choose which citizens they will respond to.

Cory Wanless weighed in next on the matter of regulatory negligence. "The ERCB knew about problems with Encana's drilling in Rosebud," Wanless said. "They knew, for example, that Encana was hydraulically fracturing at shallow depths in close proximity to the Ernst well. They knew that there were serious problems with the regional aquifer, including concerns about potential contamination with well water. They also had been informed that the radical change in Ms. Ernst's water may have been linked to Encana's unconventional activities. And the ERCB also knew that Encana had perforated and fractured directly into the Rosebud aquifer. And despite all this knowledge, the pleadings say, the ERCB failed to conduct any form of investigation whatsoever and failed to respond in accordance with its investigation and enforcement process. In the building inspection cases, the important issue is whether the government agency has decided to implement an inspection scheme. Once it has, it cannot then negligently fail to inspect."

Later that afternoon, Nancy McCurdy, a lawyer acting for Alberta Environment, informed Judge Veldhuis that Ernst's shorter statement of claim still needed a thorough vetting. The claim contained not only "substantive and improper defects," said McCurdy, but "misleading" words such as *hazardous, pollutants,* and *contaminated.* According to McCurdy, any reference in Ernst's claim to other landowners with contaminated aquifers was improper too. Moreover, any allegations regarding how the oil and gas industry operated didn't belong in the lawsuit. Accounts of diversions from freshwater aquifers without permits and other breaches of the Water Act should be struck from the claim, McCurdy argued. Ernst's lawsuit was simply about what had happened to one person and "not by its very nature a public inquiry into fracking or into the oil and gas activities in Rosebud."

Klippenstein offered a short rebuttal to the government palaver. He explained that the case wasn't a car accident tort but dealt with groundwater, a public resource that traveled great distances underground. Contamination in that water, "almost by definition," would move and affect other landowners. Words such as "hazardous" and "pollutants," he said, merely described the point of the case. "My friend may not like that her client is being sued for contamination or having a role in it, but that's the fact... Ms. Ernst should not be unduly muzzled to satisfy the tender delicacies of the defendants."

At the end of the day, Judge Veldhuis looked ruefully at two bankers' boxes full of documents and admitted she had "a really huge task ahead" of her. She thanked the crowd in the courtroom for their courtesy. She also promised to deliver a decision "as soon as I can."

But before Justice Veldhuis could rule on whether the ERCB had immunity from civil action, the federal government removed her from the case. Less than a month after the January hearing, the Harper government promoted Justice Veldhuis to the Alberta Court of Appeal. The promotion didn't worry Klippenstein at first. In such cases, he told Ernst, judges usually ruled on their active cases before moving on. However, in a conference call with the lawyers, Justice Veldhuis revealed that such a courtesy in this case was "not an option" due to the immediacy of her appointment. The news stunned Ernst's lawyers, though Ernst had predicted all along that the Harper government would try to interfere in the case. The federal government, an active supporter of hydraulic fracturing, had now moved the case back to square one. Although the decision meant months of delay and additional costs, Ernst stood unfazed: "I won't be intimidated by the abuse of power."

Two weeks after the Veldhuis appointment left Ernst's case in legal limbo, Klippenstein alerted his client to another curious development. Alberta Court of Queen's Bench Chief Justice Neil Wittmann, who had been appointed by Harper in 2009, had volunteered to take over the case. That left Ernst with three options: the three defendants, including Encana, could sit down with Ernst and come to an agreement on a new judge; her lawyers could "roll the dice" and accept whatever new judge the system might select or Ernst could accept Wittmann's out-of-the-blue offer. Klippenstein said that Wittmann's offer was hard to read. But Ernst regarded all the choices as bad news. Given that negotiating with the defendants would cost her more money and time, Ernst instructed her lawyers to accept Wittmann as the new case manager. Rather than hold another hearing, Wittmann advised all the parties involved, he would read the transcripts and the court briefs. Then he would make a ruling on whether or not the ERCB could be sued.

As Wittmann deliberated and months passed, debates about fracking heightened throughout the province. In March, the Alberta Association of Municipal Districts and Counties, which represents much of rural Alberta, voted overwhelmingly in favor of a resolution calling for stricter regulation of hydraulic fracturing and for higher levels of protection for groundwater. The AAMDC resolution demanded that the government directly map all groundwater and its contents, perform pre- and post-seismic testing on frack jobs, and "protect surface and groundwater supply by imposing a minimum wellbore casing depth below aquifer zones."

Wheatland County Councillor Brenda Knight had introduced the resolution after gathering evidence on several hydraulic fracturing incidents in the county where Ernst lived. The damage included groundwater contamination, water-well losses, improper well casing, and large sinkholes that appeared unexpectedly in the ground. In one case, a farmer lost his tractor in a spring that opened up. The practice of hydraulic fracking had been around for a long time, Knight acknowledged. "But it used to be a Model-T, now it's a Lamborghini." Protecting landowners and the environment was not "a rocket science," she added. "Groundwater is the source of life for our household and our cattle. We can't ruin that."

In April 2013, Jessica Ernst watched a disturbing television documentary by the Australian Broadcasting Corporation, *Gas Leak!*, on the fracking of coal seams under agricultural land in Queensland, Australia. The report showed footage of novel seeps of methane bubbling up in the Condamine River. Disgruntled farmers complained about dropping water tables. For the camera, they also lit their contaminated water on fire. One citizen asked if coal fracking wasn't "this generation's asbestos." The documentary also followed a group of researchers from Southern Cross University as they measured gas levels in the atmosphere above fracked and unfracked landscapes.

The researchers found that methane levels and, later, radon and CO_2 levels were three times higher in fracked geographies. They asked why, prior to the drilling, no one had collected basic baseline data on gas migration in the air or water.

As Ernst clutched her chest, the documentary answered that last question by telling the story of Simone Marsh, an "environmental specialist" and a single mom. Like Ernst, Marsh was an industry insider. For nearly twenty years, the Brisbane resident had worked on the environmental approval of mining projects in Queensland's resource industry. In 2010, the Queensland government asked Marsh to review environmental impact assessments on two coal-seam fracking projects worth nearly $50 billion. Although the projects promised to puncture the landscape with 40,000 wells and associated pipelines, all for natural gas export, Marsh was given no baseline data. However, her supervisors, two of whom later went to work for the coal-fracking industry, did give Marsh very specific instructions: "I was taken into a meeting room, sat down and told that there wasn't going to be a chapter on groundwater, and I was... stunned. I said, 'What are you talking about? What do you mean there's not going to be a chapter on groundwater? It's one of the biggest issues for the project.'" In a series of emails, Marsh later advised her superiors that they were breaking the law. They ignored her and approved the projects anyway. A nine-hundred-page freedom of information request by the Australian national broadcaster confirmed that the approvals process for fracking coals was "rushed, made with insufficient information and put commercial considerations above environmental ones."

The Australian documentary made Ernst feel like she was watching a horrific rerun of what she had experienced in

Alberta. Another dishonest government, an ocean away, had refused to collect baseline data before brute force was applied to fractured geologies. The women bore the same industry title, "environmental specialist," and even wore the same scarf. Ernst researched Marsh's story further after the airing of the documentary and found it crowded with gatekeepers. The whistleblower testified before Queensland's Crime and Misconduct Commission, saying the rapid approval of the two projects clearly violated the Environmental Protection Act. The commission, however, had forgotten to tell Marsh that it had no jurisdiction over environmental assessments and couldn't do an investigation. It therefore concluded that there was "no evidence of official misconduct." Disgusted by the corruption, the soft-spoken Marsh quit the business and sold her house so she could go back to university and retrain as a psychologist. "I've lost complete trust in the government and the industry. I can't work with either of them," she later told a reporter. Throughout the affair, the industry had dragged out its standard defense: blame nature.

Back in Alberta, the interests of industry and government had increasingly become indistinguishable. In April 2013, the Alberta government appointed Gerard Protti, former Encana vice president and a founder of the Canadian Association of Petroleum Producers, to chair the newly formed Alberta Energy Regulator (AER). At the time of his appointment, Protti was still listed in Ottawa as a registered energy lobbyist. He also sat on the board of directors of the Alberta Research Council (now Alberta Innovates), which had dismissed water contamination cases as natural.

The Alberta government had first hired Protti to provide advice on streamlining legislation to make the extraction of extreme hydrocarbons easier. They liked the energy lobbyist's recommendations so much that they put him in charge of implementing the changes. One of Protti's first suggestions was that the province draft a new energy bill. The controversial Responsible Energy Development Act disbanded the scandal-ridden ERCB and created the generic-sounding Alberta Energy Regulator, now 100 percent funded by industry. In addition, the bill transferred environmental and water protection from Alberta Environment to the energy regulator. To discourage public litigation, the new act also gave the AER authority to limit public involvement in industry hearings. And to foster the speedy approval of unconventional projects, the words "public interest" were removed from the agency's mandate. Not surprisingly, the act also sported a rewritten immunity clause that included the phrase "any act or thing done or omitted to be done in good faith under this Act." Lawyers dubbed it the "Ernst clause," because Ernst's lawsuit emphasized what the ERCB had promised but failed to do. It was now impossible for any other Albertan to sue the regulator for failing to protect the public interest.

Protti himself was no stranger to Ernst's case. While working for Encana, he had stalwartly defended the company's abrupt invasion of Rosebud in 2005, arguing that Encana had "gone the extra mile to ensure that plans are clear." A year after Encana fracked the Rosebud aquifer, Protti claimed that "the targeted CBM formations do not produce water," even though many of them did. When asked about Jessica Ernst's burning water in 2006 by a university magazine, the executive had replied that Encana "can demonstrate that we've never impacted an aquifer in our drilling." He added that Ernst's flaming water was "fairly common" in areas that produce natural gas and that the bubbly water was safe to drink.

Protti's appointment struck Ernst and many others as a

farce: the former executive of a company that pushed fracking and blamed methane water contamination on nature now
controlled the regulation of water permits as well as oil and
gas activity. Even industry cheerleaders compared the appointment to putting a fox in charge of the hen house.

In August 2013, one year after being served Ernst's shorter
statement of claim, Encana finally filed a Statement of Defense,
and even then only because Justice Wittmann asked for it. The
document consisted of only six pages. Predictably, the company swore that it had complied with or exceeded all relevant
laws, regulations, and directives, and that the methane contamination of groundwater "occurred naturally or by other causes."
And even though company gas well histories and logs clearly
showed Encana perforating and fracking the 05-14 well, the
company now argued that the gas wells in question had been
"stimulated" as opposed to "hydraulically fractured." Company
lawyers described stimulation as "pumping inert nitrogen gas
into an individual coal seam at a high rate."

Encana's defense also claimed that the company had
offered to test Ernst's water well but that Ernst had blocked
access. Ernst laughed at that lie. Had the company forgotten
that it tested her water well in 2003 and gave it a "clear" rating
for methane? Although the company had offered to test Ernst's
water well a second time, their offer included a strict deadline
for acceptance. Encana mailed the offer to Ernst twenty-four
days after that deadline expired.

At the end of its defense, Encana hinted that Ernst might be
crazy or unstable. "Encana states that to the extent the Plaintiff
suffered any personal injury, which is denied, any such injuries
or problems are unrelated to the Operations and are the result
of the pre-existing medical, psychological, or other conditions

of the Plaintiff." Interestingly, Ernst's lawsuit contained not one word about health or, for that matter, personal injury.

IN SEPTEMBER 2013, Jessica Ernst made what she calls "a fate- 263
ful, terrible trip" to Newfoundland. Magic's health was failing, and she did not want to leave him again. In 2012, her dog had spent two and a half months in the kennel while Ernst told her story in Ireland, Michigan, Nova Scotia, and New York. But Black Spruce Exploration wanted to frack the rugged island just outside the boundaries of Gros Morne National Park, a UNESCO world heritage site. Community groups pleaded for her to come, and Ernst relented. One night four hundred people came out in Stephenville, on the province's west coast, to hear the environmental scientist speak. Ernst told them that clean air and water were irreplaceable: "You have something here that is worth billions," she said, and worth much more than any amount of oil. One resident talked of the stark beauty of the place and described the proposed fracking projects as "a turd in the middle of a china plate."

While Ernst talked to rural Newfoundlanders, Brent O'Neil, an international oil-patch driller, sat down in Calgary with lawyer Glenn Solomon to get some legal advice. O'Neil's mother, Ann Craft, owned an eighty-acre farm near Ponoka, Alberta, and in 2012 her land got fracked. When Quicksilver Resources worked over four shallow coal seams north of Craft's home, a fracture went out of zone, lifted up her front porch, and buckled several outbuildings. Alberta Environment had promised to investigate all the wells drilled within a two-kilometer radius, and also to catalogue the fracking fluids used. It had even ordered a hydrogeological study. But Quicksilver talked the regulator out of such rigor, and no proper investigation ever took place. To add insult to injury, a trucking company delivered tainted produced water to Craft's cistern. She bathed in the toxic brew, and it nearly killed her.

After two years of wrangling with the government, Brent O'Neil wanted to know what chances his mother might have if she sued the regulators for negligence. O'Neil taped his conversation with Solomon on his iPhone, because he planned to share it with his mother later.

"Take a step back," Solomon advised O'Neil in fatherly tones. "I told you on the phone, I act for ERCB when they're sued on these types of things. There's only one such case in Alberta that I'm aware of where they're using outside counsel, which is me at the moment. And that's an oil spill out in the Rosebud area, which has become more of a political grandstanding issue than a legal dispute."

"Over an oil spill?" asked O'Neil for clarification.

"This was a fracking case," Solomon replied.

"Oh," said O'Neil.

"It was alleged contamination of a water well. Doesn't appear to be any personal injuries. And..."

"Just groundwater contamination?" interjected O'Neil.

"Groundwater contamination," confirmed Solomon. He continued: "Encana is the oil company. They've said, 'We deny that we've done anything but we'll give you a lifetime supply of potable water anyway, because we just don't care and we don't want to fight with you.' You know, it's Encana, and they have all the money in the world. And Alberta Environment and ERCB have been sued in that one as well. I can tell you it's a case that is seven years old. I haven't yet filed a Statement of Defense because it's been tied up in preliminary applications... because that's what happens when you start suing Alberta Environment and ERCB."

Solomon went on: "We keep on telling the plaintiff's law-yers, look, if you get rid of us [the dispute with the regulators], Encana is going to resolve this with you, cause they always do. That's what they do. Encana has said, 'Look, you know, we're happy to pay for this, without admitting or denying liability... You know, it's... this is a rounding error on our balance sheet, for God's sakes. Would you stop being a nuisance?' "

265

"But the PR and the bad publicity that comes from it for everybody, is that even worth it?" asked O'Neil.

"Encana, ERCB, and Alberta Environment just don't care about that either," responded Solomon. "They just don't care about bad publicity because... what tends to happen is that the people who go yapping to the media are typically seen as nutcases."

O'Neil then asked a direct question. "On your experience with fracking and stuff, where, what's the success rate?" O'Neil noted that Quicksilver had had a claim filed against them by Dale Zimmerman, the Wetaskiwin farmer, involving fracking and groundwater contamination. "What's the Canadian cli-mate for that kind of stuff? Is it worth a fight?"

"I'm not aware of any cases that have gone to trial where fracking damage has been successfully proved," Solomon replied. "But, again, most of these cases resolve. 'Okay, we damaged your water well. We'll just set you up with potable water through a tank system forever, because, you know, we just spent a million dollars drilling this well that we made a hundred million on. And it's costing us an extra three hundred thousand. We're okay.' "

Solomon elaborated on the industry's attitude: " 'You know, we don't need to litigate with you, we don't even need to know that it was our fault. We're just happy to pay you. And by the way, by doing that you shut up, the regulators stay off our back, we get to do it again down the street.' And so that's the oil company approach on these [things]. The people who typically are suing are getting a lot of resistance, and it's a knock-'em-down, drag-'em-out brawl, where the oil companies are not resolving it. If you drag in the regulators, I can tell you from experience... it's World War III. And Encana, Alberta Environment, and the ERCB, as it turns out, all have effectively unlimited resources. You know they have office towers full of experts. They have bank accounts full of cash. The cost of having even an army of lawyers is something that they wouldn't even notice, and they don't have to answer for it. So anyone who wants to pick that fight literally is crazy."

O'Neil interjected, "Yeah, it's almost—it is, it's terrifying as a landowner in Alberta, like, to see what my mom's gone through, and as you say, what she has to fight, or potentially look forward to fighting, it's—it's so scary."

"It is scary, and it's expensive," confirmed Solomon.

O'Neil left the meeting bewildered. He was struck by Solomon's cockiness, and he couldn't believe that such a prominent lawyer would be so matter-of-fact about how industry silenced fracking liabilities with sealed settlements. Based on Solomon's advice, Ann Craft did not sue the regulators. She simply sued the company that had delivered the toxic water to her cistern.

Brent O'Neil sent Jessica Ernst a copy of the audio several weeks later, but he didn't release it to the media for another year. When a transcript went public in 2014, citizens reacted angrily to Solomon's reasoning about how frackers played the legal system. Quoted in *The Tyee*, an online newspaper, one Alberta mother said Solomon's statements raised a few big questions: "Why does this lawyer's statement, 'What tends to happen is that the people who go yapping to the media are typically seen as nutcases,' sound eerily reminiscent of something a pedophile would say? 'This will be our little secret, and if you try and tell, no one will believe you.' (I am not calling this lawyer a pedophile, merely comparing what I think are similar intimidating, belittling and manipulative techniques.)"

267

The mother posed another critical question: "What are we teaching our kids by signing away THEIR right to freedom of speech? Contaminated water is a public health issue. Water moves, even contaminated water. Gag orders should be illegal."

DAYS AFTER ERNST returned from Newfoundland, and seven months after the Harper government had removed Justice Veldhuis from the case, Ernst's lawyers found Chief Justice Wittmann's ruling on the Internet. Just as Ernst had predicted, the decision granted the ERCB immunity from civil action, citing the regulator's statutory immunity clause. Wittmann did, however, express some sympathy for the Charter argument: "To a certain extent, a claim for a Charter breach is based upon the establishment of a right and an infringement of it by the action of a government or government agency. That is what is alleged here and, however novel the claim might be, I cannot say that it is doomed to fail or that the claim does not disclose a cause of action."

Wittmann dismissed the slanderous statements about Ernst being an eco-terrorist. But ERCB could not rely on its argument on the Wiebo Ludwig eco-terrorism claim, he found because, in terms of evidence, "There is none." He also threw out the application from Alberta Environment to censor and rewrite Ernst's statement of claim by removing all mention of contamination and fracking: "Tinkering with pleadings by a Court is not, in this case, useful to the advancement of the action, in accordance with the foundational rules." He then ordered the rest of the lawsuit against Encana and the Alberta government to proceed.

Ernst didn't take long to reflect on the matter: she wasn't about to let the courts and Stephen Harper use her case to cripple the Charter. She ordered her lawyers to appeal. "I have no choice," she told the media. "Chief Justice Wittmann ruled that the ERCB has a duty to protect the public, but not me. I am the public; we all are. Without water to bathe in, the public's well-being declines—as mine has for years. Without water to drink, the public dies, so do all individuals—including Justice Wittmann and his loved ones. There is no frack worth that. If I let these rulings stand, companies and the Canadian Association of Petroleum Producers will push regulators across Canada to do unto others as the ERCB did unto me, when they breached my Charter rights and tried to intimidate me into silence by judging me a criminal without any evidence or trial."

On November 19, 2013, after suffering a series of strokes, Bandit's quiet brother, Magic, died. Ernst felt eternally guilty about kenneling him so often while she traveled to give talks. "He died lying beside me, at noon," wrote Ernst to friends, "me holding onto his heart, as he always held onto mine, even when I never knew it." The lawsuit had now outlasted both her dogs.

Once Wittmann had granted the ERCB immunity from civil action, Alberta Environment decided to employ the same argument. Having failed to censor words like "pollution" from the claim, government lawyers demanded in a February 18, 2014, filing, that the court strike Ernst's entire claim, arguing that Alberta Environment owed no duty of care to landowners. Copying Glenn Solomon's arguments almost verbatim, lawyers for Alberta Environment explained that the spirit of provincial environmental legislation did not protect land-owners from harm or injury and that all litigation should be directed against companies: "The overall tenor of the Environmental Protection and Enhancement Act does not create a responsibility owed by the Province to those who suffer the consequences of escape [of air or water pollution], but rather to try and ascertain who caused escape, and where it can be shown that an entity or person caused escape, to ensure that the person or entity responsible for the cause of the escape remediates the effects of that escape." In other words, the legislation was "permissive, providing the Province with the discretion of when and how to take action or no action as it sees fit in a particular situation."

According to the testy government brief, Ernst was an unreasonable woman who wanted to dictate to the government how to do its water-contamination investigations. "What the Plaintiff's claim is seeking to do is to put her at the table with government workers and dictate to them the way they should properly go about their business... Ms. Ernst does not want fracturing to occur, and she wants to have a say in how testing should be conducted. This would be an infringement on the discretion afforded the Province's agents under the EPEA, and the statutory immunity that flows from the use of that discretion." Furthermore, the government said, the Ernst case raised the issue of indeterminate liability, "as there are thousands of energy and other commercial projects situated in this Province, any one of which may have environmental impacts on surrounding landowners."

The government brief also played what it thought was the ultimate trump card: the Alberta Research Council report. It explained, "The Council used a variety of evidence...in reaching the conclusion that the methane in the Ernst well was biogenic and that energy development projects in the area had most likely not adversely affected the Plaintiff's water supply. The Council was not only aware of sampling contamination issues and errors but factored this into its independent analysis. Based on the independent findings of the Alberta Research Council, the Province closed its investigation." Therefore, Ernst's lawsuit had no legal basis whatsoever.

In their brief, Ernst's lawyers argued that the case was about "bad faith" and reckless actions by the government that had resulted in methane and hydrocarbon contamination of Jessica Ernst's water well. Moreover, the Klippenstein brief said, Alberta Environment's investigation "was ad hoc, irrational and beset by serious errors." The brief detailed the incompetencies witnessed by Ernst and other landowners: "There was no sampling protocol. Samples were contaminated. Alberta Environment lost or destroyed data its investigators had collected. Investigators entirely failed to investigate specifically identified Encana gas wells that had been fracked either directly into or near the Rosebud aquifer. Investigators failed to obtain from Encana a list of chemicals used in its fracking operations, and correspondingly failed to test the water for possible 'red flag' contaminants that would help identify the source of the water contamination."

The firm also argued that Alberta Environment's application to strike the lawsuit constituted "an abuse of process" designed to cause unnecessary delays and to exhaust Ernst's financial resources. "This behaviour should not be condoned and should have cost consequences," the firm rebutted.

Four days before the Alberta Environment hearing, Cory

Wanless remembered seeing, in Ernst's heavily redacted FOIPed materials, some emails about the government editing the Alberta Research Council report. Wanless emailed Ernst and explained that they might be able to prove ARC wasn't independent, as the government now claimed. Ernst cheerfully replied, "Hey, I got them unredacted." Over the next two days, Ernst found and scanned a collection of government correspondence that proved the ARC acted as a secretary for Alberta Environment. Her lawyers sent the evidence to government lawyer Neil Boyle, two days before the hearing.

Chief Justice Wittmann heard the government's unprecedented motion to throw out the lawsuit on April 18, 2014, in the Drumheller courthouse, the legal jurisdiction where it belonged. More than forty people attended, including nearly a dozen Rosebud residents, a development Ernst found immensely healing. The Alberta government lawyer, Neil Boyle, warned that if the judge didn't strike down the Ernst case, it could open a floodgate of litigation against the province, with potentially "millions or billions of dollars worth of damages." Although Boyle emphasized that the immunity clauses in the Environmental Protection and Enhancement Act and the Water Act shielded the province from any civil action, questions from Wittmann made it clear that these immunity clauses in the acts differed substantially from the ERCB's Section 43.

Boyle did not bring up the Alberta Research Council report, because the documents Ernst's lawyers had sent him showed that Alberta Environment had heavily edited the report it contracted. No one in government had apparently warned Judge Wittmann, however, who eagerly asked about the report in court.

Wittmann's question gave Klippenstein the opportunity to file the damning documents as court evidence. That was a highly unusual turn of events during a motion hearing to end the case. Wittmann allowed the evidence to be filed, and the Alberta government lost what it regarded as its get-out-of-jail-free-card. Had the government been successful in using the ARC report to quash the lawsuit, Ernst suspected that Encana would have applied to do the same. Her dogged freedom of information requests had paid off.

Later in the hearing, Cory Wanless reminded Justice Wittmann that governments can and do owe private duties of care to individuals, as well as having wider public obligations. Police, for example, he said, owe a private duty of care to an individual while investigating a crime. So, too, he argued, do provincial inspectors checking possible contaminants in water wells drawing from aquifers that sustain a third of Alberta's population.

At the end of the hearing, Wittmann announced that he would make his decision after the Court of Appeal had ruled on the ERCB and on Ernst's Charter claim. Although Encana's lawyers were not in the room that day, the chief justice "recommended" that the next phase of the lawsuit proceed with an exchange of documents. (On a later case-management call, Wittmann would order Ernst to produce all relevant records by October 31. He gave Encana an extra six weeks, until December 19. That angered Ernst.)

Some of the landowners who had attended the hearing gathered afterward over hamburgers at Drumheller's Bernie & the Boys Bistro to discuss the case. What made Ernst's legal journey so special, reflected Glenn Norman, a burley farmer from Bowden, was that "Jessica Ernst refuses to be bought out."

Norman had driven more than an hour to the court to show his support. "She has been wronged and she wants rightful justice. That's why her case has caught so much public attention." Shawn Campbell, the semiretired rancher who had himself battled Encana for years, declared, "Jessica is fighting for the welfare of the public in general. To me, the issue is government transparency, and we have none of it. They're in the deepest closet they can find." Another farmer aptly summarized the case: "Industry just pissed in the wrong bowl of cornflakes.'"

273

While preparing an appeal to Chief Justice Wittmann's ruling, Klippenstein's firm marshaled arguments that would resonate with landowners and farmers across the continent: "Ernst, like all rural landowners who live near oil and gas development, has little say in where oil and gas operations are located or how such activities are conducted. She has no ability to inspect operations, or to make sure that the operations are conducted in a safe manner, and only a limited ability to respond to protect herself or her property when something goes terribly wrong. In this case, Ernst was completely reliant on the ERCB to protect her and her property from adverse impacts caused by negligent oil and gas activities. Where citizens have no means to protect themselves from a real danger, they should be entitled to rely on government agencies tasked with inspection and enforcement."

In May, Ernst's appeal on whether immunity clauses protected government regulators finally landed before three Court of Appeal justices. The panel of judges heard the case in a blinding white, ultramodern courtroom on the twenty-sixth floor of the TransCanada Tower in downtown Calgary, just a floor above the offices of the law firm representing Encana. Before the hearing began, Ernst surprised ERCB lawyer Glenn

Solomon by shaking his hand and wishing him good luck. Thanks to Brent O'Neil's tape, she knew that she was shaking the hand of a man who regarded her as a crazy political grandstander now involved in World War III.

Landowners again packed the courtroom, which this time resembled something from the set of *Star Trek*. Laypeople had trouble following the arcane legal arguments. Incredibly, the Court of Appeal indicated a willingness to try the case on the spot but couldn't because the ERCB hadn't filed a Statement of Defense. The justices complained there wasn't enough information in Ernst's now-shortened statement of claim, and then asked questions about little-known legal precedents. Ernst's lawyer tried to steer the three middle-aged white men to the matter at hand. "No government or province can legislate themselves out of the fundamental rights guaranteed by the Charter of Rights, on which Ernst's claim is based," Klippenstein argued. He called the Charter "the supreme law of the land."

Solomon, representing the ERCB, repeated his well-rehearsed lines: the board's immunity clause exempted it from Charter claims, and the regulator hadn't caused any damages or harm to Ernst. "The damages existed when she came to us," insisted Solomon. In any case, he said, a sensible lawsuit would target only the regulated company for damages, not the regulator.

It wasn't clear when the justices would make their ruling, but given the makeup and tenor of the court, Ernst didn't expect a ruling in her favor. Outside the courtroom, though, damning evidence about fracking continued to mount. That summer, the Council of Canadian Academies released a national report on fracking that concluded the industry had grossly outpaced both science and public policy. In a Toronto talk, one of the report's key authors, famed hydrogeologist John Cherry, said, "I found no cases where rigorous groundwater monitoring has been done at any fracking pad. Exactly zero, not a single one. Anywhere, ever." Asked why government was reluctant to protect a public resource as valuable as groundwater, the hydrologist replied that it costs money to monitor past societal mistakes. "Groundwater pollution develops slowly over years and decades. If there is anything that government can shrug off to the future, it's groundwater."

At the same time, a University of Waterloo report described the nation's 500,000 leaky oil and gas wells as a significant threat "to the environment and public safety." The fluid injection of steam, water, sand, or chemicals to force out more hydrocarbons "elevates the mechanical and thermal loading on wellbores, and significantly increases the probability of leakage problem development during the operational lifetime of the wellbore, before final abandonment," the report said. Fixing badly cemented wells, it confirmed, amounted to little more than the "persistent under-reporting of negative results."

Ernst never thought she'd get a dog again. But she changed her mind that August when she heard about a ten-year-old Jack Russell named Gem, needing a home. As soon as Ernst met the black-and-white dog, she knew she had found an "adorable" companion. Within days, Gem had settled in at Ernst's bungalow and was hunting gophers. "Gemilicious" also dribbled basketballs across the prairie, making Ernst laugh. Suddenly the weight of the lawsuit seemed manageable again.

ON SEPTEMBER 15, 2014, Alberta's Court of Appeal judges ruled that the ERCB could not be sued. Their densely worded, eleven-page decision announced that the Energy Resources

Conservation Board (now the Alberta Energy Regulator) owed no duty of care to individual landowners harmed by industrial activity. Furthermore, the justices added, the immunity clause in the Energy Resources Conservation Act, Section 43, protected the regulator from any lawsuit or Charter claim, regardless of how the regulator treated individual citizens. "Provisions immunizing decision makers from liability are not so uncommon or unusual in free and democratic societies as to rend them constitutionally unreasonable," explained the ruling.

Ernst thought it was a morally stupid decision that protected her greatest abuser. "The courts only reflect what Albertans vote for, and they vote for these kinds of civil abuses as long as revenue flows from hydrocarbons," she told her lawyers. Wanless believed the ruling should alarm Albertans, as he told a reporter: "The tenor of the judgment is that the regulator knows best, and because it knows best, citizens should not challenge what it does in court. How can a provincial government grant immunity to a regulator for infringing Charter rights?"

After the Court of Appeal decision, Ernst added up the costs of seven years of resistance to government negligence. She had lost two major legal battles. The most culpable party, the ERCB, had been excluded from her lawsuit. She expected Alberta Environment to be granted immunity too. Her legal fees and expenses had claimed more than $300,000 of her savings, and the pressure of keeping significant funds in a trust account for the lawyers was unnerving. She hadn't bought any new clothes, except for socks and underwear on sale; friends had thoughtfully donated clothing for court appearances. She rarely ate out and did not go to the movies. To save gas money, she visited friends infrequently. She had lived for nine years without safe well water, and still hauled water by truck for cooking and cleaning. She composted her toilet in the yard and in the winter took a sponge bath once a week. She washed her hair every ten days.

But as the doors closed one by one, Ernst realized she didn't fear defeat, let alone the power of the gatekeepers. She recalled the words of her psychiatrist years earlier: the abuse she endured as child would someday give her the power to transcend any hardship. That day had come. Klippenstein had warned her in 2007 that though she might lose her home and everything else, "at least you will be able to walk down the street, knowing you did everything you possibly could." Were her lawyers willing to go all the way to the Supreme Court of Canada on the issue? Ernst asked. They said yes.

As Ernst saw it, Encana, the Alberta government, and the Canadian legal system had become her unwitting allies. Every time the ERCB or a government lawyer defamed her in a legal brief, she won another small victory for groundwater. Every time the court system upheld another falsehood, her case advanced the need for greater transparency and accountability. The many delays and interferences only proved that industry and government were guilty and that the frackers had broken the law. "My job is done," Ernst told a reporter. "The skeletons of hydraulic fracturing can't be put back in the closet." The longer her case dragged on, the more ordinary citizens would realize how corrupt the legal and regulatory systems supporting fracking had become. To her astonishment, her case had reached an audience far beyond Rosebud, Alberta. People from Argentina, Poland, Chile, the Netherlands, Sweden, Spain, Ireland, France, Germany, Canada, the United States, the UK, Denmark, South Africa, and Australia now followed her legal battle.

No matter what happened next, Ernst decided, it would energize her resolve. She made her application to the Supreme Court and patiently waited for a reply. As the chief justice had ordered, on October 31 she tabled 2,136 documents supporting her lawsuit. Her case might cost hundreds of thousands more in legal fees and occupy the rest of her life. The artwork and everything else she owned would have to go. She didn't expect to find justice, as Klippenstein and Wanless still hoped she would. Instead, she realized that her case had become something more profound: the public exposure of the groundwater abuse and regulatory corruption funded by the world's most powerful industry.

She also knew by now that she had a power no government could ever defeat: "Men," she said, "do not understand the courage of ordinary women."

FOURTEEN

The Sisters
of Jessica Ernst

DIANA DAUNHEIMER, AN organic vegetable farmer and
mother of two children, got her first introduction to
horizontal drilling and hydraulic fracturing in Didsbury, Alberta, in 2008. When industry fracked two wells
simultaneously, with diesel fuel, just 1,300 feet from her farm,
Daunheimer discovered that her dream home lay on top of the
Pembina Oil Field, the largest conventional oil pool in western Canada. In the 1950s, industry had extensively drilled and
fracked the field with vertical wells and thousands of water
injection wells, too. Now industry was again attacking the
Cardium Shale, the field's chief hydrocarbon-bearing rock,
with greater force: high-volume multistage horizontal fracking. The Alberta government hailed the stubbornly tight rock
as a "mini-Bakken." In three years, the number of fracked horizontal wells in the region catapulted from seventy to several
thousand.

The shale oil boom around Daunheimer's house unfolded the same way the shallow frack revolution had near Rosebud. In Alberta, frackers paid significantly reduced royalties on their first 50,000, 60,000, or 70,000 barrels of oil, depending on the well specifics. To keep things economic, self-regulation remained the rule. No provincial regulator bothered with cumulative impact assessments on communities or land. Nor did any regulator set up a proper air-quality or groundwater monitoring program. Thousands of abandoned and orphaned wells dotted the Pembina field, but none were audited for leaks or migrating gases. Instead, industry and the regulators sang the usual song about fracking being proven and safe. Two wells near Daunheimer's farm soon multiplied into six. Frackers encircled her home, garden, and family.

Daunheimer, petite and scrappy, had moved to Alberta from Ontario as a child. She traveled to Australia after university and then worked for several years as an environmental coordinator for the city of Airdrie, Alberta, before becoming a mother. Her husband, Derek, worked as a rig manager in Saskatchewan's and Alberta's heavy oil patch, so she knew a few things about the industry. Like most landowners, she had assumed that regulators did their jobs and policed the rotten apples in the barrel. But as fleets of pumping and proppant trucks buzzed through her community, she learned differently.

Her education didn't happen overnight. After the first two fracked wells left Daunheimer's family breathing toxic fumes, she wrote off the experience as the price of living in a province dependent on hydrocarbon production. But when Angle Energy fracked a sour gas well with propane just 1,237 feet from her house, Daunheimer got alarmed. Sour gas, or hydrogen sulfide, can damage the lungs and is a potent neurotoxin and memory eraser. The gas had killed scores of energy workers in the province, displaced hundreds of landowners, and injured thousands of animals. The sour frack job put Daunheimer on high alert.

Increased air pollution followed. Industry wanted only the oil in the Cardium, so it flared off all associated gases into the atmosphere. (In the Bakken in North Dakota, industry burned off $100 million of natural gas into the air every month.) For nearly three weeks, Angle Energy burned its waste gas in two incinerators. The equipment sounded like a jet engine, and at night the flames roaring from the machines blinded Daunheimer's household. While the smoke and fumes terrified her chickens and goats, the pollution gave the family chronic headaches, dizzy spells, and respiratory infections. Three weeks later after the massive gas burning, Daunheimer's goats aborted 50 percent of their offspring. Shortly afterwards, Daunheimer noticed a hard lump on her seven-year-old daughter's neck. Doctors later diagnosed the rare tumor as a calcifying nuchal fibroma and said it was too risky to operate. When Daunheimer went to local public health officials to ask about the impacts of oil and gas drilling, they told her the subject remained a "hot potato on the table and nobody wanted to touch it."

Her daughter's strange tumor, combined with the toxic fumigation, got Daunheimer reading and researching "like a madwoman." She first found one of several alarming studies by Alberta researcher Mel Strosher, completed more than fifteen years earlier. Industry flares burned so inefficiently (at only around 62 percent), the studies showed, that they spewed 150 hazardous unburned hydrocarbons into the air. The contaminants—including benzene, styrene, ethynylbenzene, ethynyl-methyl benzenes, toluene, xylenes, acenaphthylene, biphenyl, and fluorine—floated downwind into the lungs and homes of landowners. A study by U.S. environmental health expert Theo Colborn had found methylene chloride, a toxic solvent, in the air near natural gas operations, too.

The ERCB, however, assured Daunheimer that she had nothing to worry about. The regulator denied Strosher's findings and said that all flares and incinerators in Alberta burned with 99 to 100 percent efficiency. Then Daunheimer discovered that Angle Energy had been venting sour gas from a well site without proper public notification. At her insistence, the regulator investigated and found "a failure by Angle Energy to include accurate information about its plans to vent continuously at the site." A compressor station at the same wellsite released another 15,000 cubic meters of toxic gases into the air every month. Daunheimer thought the combined emissions were "obscene." A public health analysis she found online said that 75 percent of the chemicals used at drilling sites could affect the skin and eyes; that 40 to 50 percent could poison the brain, kidneys, and immune system; and that 25 percent could cause cancer. Each scientific paper she read led to another.

Although regulators told Daunheimer that industry fracked just with water or with chemicals as harmless as guar gum in ice cream, Daunheimer discovered that wasn't true, either. Slick-water fracks didn't work well in the Cardium. As a consequence, industry often conducted 100 percent hydrocarbon fracks with diesel fuel, light aromatic solvent naptha, or kerosene. Other wells by her home had been blasted with "frac oil," synthetic frac oil or diesel invert. Daunheimer searched and found the Material Safety Data Sheets (MSDSs) on the three substances. The MSDSs, which list a chemical's health dangers, described "frac oil" as a highly flammable product that contained "benzene, a proven human carcinogen." The MSDS on diesel fuel revealed that it contained benzene, toluene, and xylene. All three toxins could foul groundwater and, if ingested, cause cancer or kidney damage and harm the brain. To Daun-   283 heimer, it didn't seem right to run these kind of chemicals through an aquifer, let alone a rural airshed.

Many fracking cocktails, Daunheimer learned, also included ethylene glycol and formamide, which were suspected of causing abnormal cell growth in fetuses. Scientists had classed other frack fluids as mutagens, because they could change DNA. Several could cross the placenta, she read, "raising the possibility of fetal exposure to these and other pollutants resulting from natural gas extraction." Approximately one hundred of the eight hundred chemicals used by the fracking industry were also so-called gender benders, meaning they could change or disrupt hormone functions in both oil and gas workers and young children. Benzene can also cause sperm mutations in men.

Meanwhile, the smashing of Cardium rock for oil in central Alberta continued. In 2012, a newly fracked wellsite just northeast of Daunheimer's home sported several unlined pits full of drilling and fracking waste. Daunheimer suspected that the poorly constructed sump pits were illegal, because they could leach water contaminated with hydrocarbons into groundwater or into the nearby Rosebud River. (Daunheimer didn't know it then, but she lived a hundred miles upstream from Jessica Ernst.) When she complained, Alberta Environment expressed some concern but said it couldn't do anything.

The determined mother then took her concerns to the newly formed Alberta Energy Regulator. An AER inspector eventually checked out the unlined pits, and his report noted that they smelled "heavily of hydrocarbon based drilling fluids." As a consequence, the inspector issued "a high risk compliance order" against Angle Energy to clean up the contaminated site.

But neither AER nor the company would give Daunheimer a copy of the clean-up order or the sampling report: she had to file a freedom of information request and pay to get the document. It showed that the AER had shut down the company for a day in 2012 while the firm removed the oily waste from just one pit. It took Angle Energy a full year to clean up the rest of the site. "If your neighbor dumped 100,000 liters of diesel oil in your backyard in the city, the fire department would be there immediately," Daunheimer points out. But there appeared to be another law for oil and gas companies in rural Alberta.

By then, Daunheimer feared not only for her children's health but for her groundwater. The family's water well stood about 10 yards from the house and was only 160 feet deep. It had been gas-free when they bought the property, but the water now tasted fishy, and her animals wouldn't drink it. Her water well was surrounded by six fracked wells, one abandoned well, and a collection of unlined pits that contained diesel fuel. Even industry warned in their literature that the Cardium was a tricky formation to frack; it lay near the Rockies, where "faults may serve as conduits for large volumes of fluid, acting as a thief zone and redirecting fluid and proppant away from the treatment zone and wasting valuable time and money."

Daunheimer checked industry records and discovered that one of the fracked wells near her home was leaking methane. The well's owner had failed to test or report the seepage to the regulator for three years. A 2010 Society of Petroleum Engineers paper explained to Daunheimer the industry's blasé

attitude about the health hazard: "The repair of these situations is a non-revenue generating exercise with the potential
to reach significant expenditures." By now, Daunheimer also
knew that horizontal wells leaked more frequently than conventional ones.

Given these concerns, in 2012 Daunheimer requested water
testing from the Alberta Energy Regulator. The AER told her
to try Alberta Environment. "I became a ping-pong," recalls
Daunheimer. Alberta Environment said it couldn't do anything because AER now governed the Water Act. A year went
by. After repeated phone calls and emails on Daunheimer's
part, Alberta Environment eventually turned up at her property. The inspectors acted pretty much the same way they had
at Jessica Ernst's place in 2006. After failing to find any evidence that the Daunheimers kept their water well in a poor
state, the inspectors asked to collect water samples from the
kitchen tap. Why in heaven they would do that, asked Daunheimer, knowing that there was a carbon filter in the kitchen?
Why not sample directly from the head of the water well? In
the end, the government purged the well for fifteen minutes,
which any expert knows isn't long enough to draw up contaminants that might be flowing through the aquifer.

A month later, Alberta Environment emailed Daunheimer
that the results looked good. The tests on her water showed
only low levels of phthalates or plasticizers. These probably
came from her plastic water pipe, speculated the government
even though her system had no plastic piping. (Phthalates,
however, have been found in contaminated wells where
industry has fracked rock.) "I will be closing the file," the government inspector's email said. "I hope this helps put you and
your family's mind at ease."

Alarmed by the government's unprofessional conduct, Daunheimer asked for a copy of the water-testing report. Instead, she got a mishmash of paper with five pages missing. Even so, she was horrified. The results indicated that some test samples had sat on the shelf for twenty-eight days, violating specific hold times. No preservative had been added to the sample for mercury testing. The government hadn't tested for hydrocarbons, either. Undaunted, Daunheimer asked more questions. Did the government have a standard list of contaminants it tested for in groundwater contamination cases? No, replied the government: testing was an ad hoc business subject to constant change.

Daunheimer bombarded Alberta Environment with more phone calls and emails. One month after declaring her water good, the government admitted that its sampling had been deeply flawed and that Daunheimer's results had got mixed up with testing for another client. "Since there were so many mistakes made," an inspector admitted in an email, Alberta Environment "will not use any of those results moving forward on the file for they would not hold up to scrutiny." At that point, Daunheimer wondered if the province's entire groundwater-testing program wasn't fraudulent. She knew one sure thing: she was on her own.

In December 2013, Daunheimer sued Angle Energy (the company had by now amalgamated with Bellatrix Exploration) for $13 million, roughly the cost of fracking three oil wells. The suit alleged negligence, nuisance, deceit, and trespass. Daunheimer decided to handle the "extremely frustrating and life-altering lawsuit" on her own, because she knew more than most lawyers about the industry and couldn't afford to do it any other way. She calls herself a mommy bear: "I protect my young, and I love where I live. I put my heart and soul into this house. I got married here. Why should I get pushed out of my home just because someone is doing stuff illegally? It's insane." As soon as she filed the lawsuit, the AER closed its files on her case.

Several weeks later, Daunheimer called Jessica Ernst and said, "I hope you don't mind, but I'm plagiarizing your statement of claim." Ernst laughed. She had read a report about Daunheimer's lawsuit in a rural newspaper and had written to friends to say, "I've been waiting for this for years." She told Daunheimer to borrow away and explained why she had made all the documents on her own case public.

Daunheimer's statement of claim highlighted the same issues Ernst's had more than seven years earlier:. "The AER investigator... remained robotic, cynical and misleading with dealing with the Plaintiff throughout her case... The directives of the AER do not adequately protect the public or the environment from harm and the ratio of inspectors to activity is so ill proportioned that compliance cannot be assured."

In its Statement of Defense, Angle Energy denied that any of its wells had contributed to injuries or damage. The company also defended an energy regulator now 100 percent funded by industry. According to Angle Energy, "Allegations in the Statement of Claim that are critical of the Alberta Energy Regulator" and AER's investigators and directives, as well as comments that the AER is "funded and controlled by the oil and gas industry," were false. Daunheimer found the comments absurd. She later attended two of Ernst's legal hearings. "Sometimes," she says, "it scares me what they have done to her and what they want to do to all of us."

A HALF-HOUR'S DRIVE north of Diana Daunheimer's place, another Alberta mother, Kimberly Mildenstein, was taking a different stand against the frackers. Mildenstein is the opposite in temperament to the feisty Daunheimer. A quiet, athletic daughter of local farmers and the mother of three boys, she coaches and referees children's hockey, soccer, and baseball leagues near Sundre, Alberta. Until the frackers arrived, the former social worker had never protested anything. But after a water truck nearly killed a neighbor who was driving home from work late one night, Mildenstein signed petitions and made phone calls to the police requesting safer roads. She felt it was her community duty.

But the industrial traffic and the speeding increased, and soon Mildenstein's neighbors were sharing stories at school about being run off the road by semi tankers supplying frack sites. Mildenstein could no long walk or bicycle on the narrow country roads with her children. Fleets of semi trailers roared past her home on Eagle Hill. The trucks put on their Jake brakes as they descended, and the helicopter-like noise thudded through the valley for miles. (A Jacobs Engine Brake sits in a box on top of a diesel engine and uses air compression to slow the vehicle down.) To protect school buses on the roads, Mildenstein asked Mountain View County to put up "No E-Brake" signs and to post lower speed limits. When nothing happened, she posted her own homemade sign. It slowed the truckers, but the county told Mildenstein to take it down.

By 2011, fracking convoys dominated the local roads. Gravel trucks, water trucks, pumping trucks, and semi trailers churned up dust and rocks. Valley residents could no longer leave the windows of their houses open due to the dust and noise. Mildenstein's own house smelled "like it had been hooked up to a muffler system," she says. To make things worse, the valley amplified the sound of flares, e-brakes, frack jobs, and incinerators.

The racket, complained Mildenstein in one letter to author-   289 ities, was "comparable to the sound of large jet engines at an airport or a helicopter hovering over the house." The ERCB sent an official out to Mildenstein's home to investigate. After confirming that drilling operations were indeed violating noise guidelines, the ERCB employee recommended that Mildenstein install a household fan to camouflage the industrial clamor. Another official with the board admitted that the industry was largely self-regulated, and that's just the way it was.

Unbeknownst to Mildenstein, the federal government around this time began to label opponents of rapid energy developments as "radicals." A special RCMP squad was set up to deal with so-called terrorism threats in the oil patch. The Alberta government followed suit. The job of the Alberta Security and Strategic Intelligence Support Team (ASSIST) was to "manage security information and intelligence" as well as to "develop threat assessments" regarding critical infrastructure, including oil and gas facilities.

As the industrial activity increased, Mildenstein's family could no longer sleep at night. To deal with the stress, Mildenstein made visits to a psychologist in 2011 and 2012; that professional recommended that Mildenstein compose more letters to the ERCB and the county about the "nightmare" of living in a fracked landscape. Mildenstein wrote and sent dozens of letters, but it didn't ease her anxiety.

In desperation, Mildenstein finally reached out to the Sundre Petroleum Operators Group (SPOG). SPOG's website described the group as "a catalyst for cohesive community spirit" whose "services are available to any stake-holder within the SPOG area with an interest in industrial development, primarily oil and gas." Once run by volunteers, the group had since been synergized; it now operated as part of a joint industry and government network partly funded by the Canadian Association of Petroleum Producers.

Synergy groups had a specific mission in the province: to deny, dismiss, and deflect any opposition to industry activities. They did so by organizing free community barbecues (more beef on a bun) accompanied by endless talks and promises about the benefits of best practices. The synergizers promoted best practices because, unlike regulations, they are totally unenforceable.

Mildenstein made a presentation to SPOG in October 2011 documenting the issues. "One truck per minute per hour— that's what our neighbor calculated," Mildenstein recalls. "That was for four months straight, day and night. You can imagine how hard it was to sleep with the Jake brakes going into the valley."

The ERCB had instructed Mildenstein to fill out complaint forms called "Objection to an Energy Resource Project." But Mildenstein, like most landowners, had trouble finding the forms on the regulator's unfriendly website. A board official who had gone to school with Mildenstein admitted that the ERCB made it difficult on purpose and helped her to print some. He also confided that he couldn't take up any of her concerns, because he needed to keep his job.

Mildenstein filed one objection form after another while industry drilled and fracked nearly fifty wells within a three-mile radius of her home. One of her objections read, "The high oil and gas traffic volumes and speeds are placing River Valley school buses at danger, our kids are going to be killed." In another she pleaded for better signage and speed limits: "This is very serious and we are in extreme danger." An ERCB bureaucrat responded by visiting her home. The regulator's first comment was to admire the new oil derrick that could be seen from Mildenstein's living room. "I thought that was kind of strange," says Mildenstein. "I realized then it was all a sham." Paddy Munro, then reeve of Mountain View County, recalls that the frackers had hit the region "like a bomb." Nobody was prepared for the traffic, the air pollution, the water demand, or the wear and tear on public roads. The ERCB acted like a tackle block for industry, says Munro. "They had been given directions that this is going to happen and to get in there and shut all the complainers down." The board "just flat out lied to Mildenstein," recalls Munro, and "SPOG was a joke."

Mildenstein persisted. She phoned the police, too, about the dangerous traffic levels. But the RCMP detachments in Olds and Sundre said they didn't have the manpower to do anything about it. A provincial peace officer agreed with Mildenstein that industry traffic posed real hazards to citizens. He told her that he had given out numerous speeding tickets but the Crown prosecutor was just throwing them out of court. Meanwhile, Alberta Transportation said it couldn't do anything because the issue was a municipal one. The municipal road officer explained to Mildenstein that the roads existed "for the transfer of goods and services." He suggested her concerns about community safety were a "conflict of interest" with that mandate.

Sleep-deprived and dogged by anxiety, Mildenstein often found her vehicle being tailgated by water trucks. The trucks followed so closely that neither she nor her neighbors could safely turn into their own driveways. "The tailgating was stalking, and it was violent," recalls Mildenstein. When she asked local farmers what they did in such situations, they replied that they followed the industry vehicles until they stopped, then gave the drivers shit. Mildenstein discovered that most trucks didn't have a phone number or company name on their doors, making careless drivers impossible to report. Whenever she drove her boys to hockey or baseball games, she now felt that she "was on the autobahn in Germany."

In early spring of 2012, Mildenstein attended a SPOG meeting on industry impacts. She watched in dismay as a local farmer presented a bottle of dirty water to the industry lobby group and said he could no longer drink or bathe in it. The farmer explained that he now had to haul his water. The group ignored the man. "I was devastated," says Mildenstein. "I had been told this was the place to go to talk about impacts, and it was a fake scene." At one point during the SPOG meeting, an ERCB official stood up and said, "We're not quite sure why we have so many objections to drilling. We think it has something to do with the movie GasLand." The comment struck Mildenstein as so darkly cynical and dishonest that she sat in shock. She realized that her requests for help had fallen on the ears of people who chose deafness.

After being nearly run off the road by a fracking truck later that week, Mildenstein received another oil-well notice. The new well was to be horizontally drilled and fracked within a two-kilometer radius of her home. That's when she snapped. Her psychologist, Sharon Comstock, would later explain in a court report that Mildenstein had breached something called a transmarginal stress threshold: "the margin or level beyond which an individual can no longer tolerate physical and or emotional suffering."

On March 15, 2012, Mildenstein penned an angry objection to the well proposed by Calgary-based NAL Energy. "Eagle Valley does not permit you to frack this location until BTX [benzene, toluene, and xylenes] water testing base line has been conducted prior to your frack and flare job. All cattle, water wells, horse tap water needs to be BTX sampled. Request frack fluid list faxed to my home. Stop Cancer," she wrote. "Concern is water aquifer contamination. You are liable for any water contamination. If you frack, I will blow up your well and shoot bullets at your crew. Now you are at risk. You are placing employees at risk." Mildenstein didn't expect anything to happen, because her other objections to the ERCB had all been ignored. She signed the objection with her family's name, phone number, and land description.

The next day, the ERCB passed the contents of Mildenstein's fax on to the Alberta Security and Intelligence Support Team and also alerted the RCMP, "due to threatening language." ASISST contacted the RCMP as well. The ERCB later explained it had reported the fax to federal police out of "concern for the safety of our staff and others who may have been at the site of the proposed operations... We cannot take threats of violence toward our staff or others lightly. Safety is paramount."

An RCMP constable was dispatched to the credit union where Kimberly Mildenstein's husband, Carsten, worked. The banker said he knew nothing about the complaint, but he explained that the level of fracking activity in the community had deeply distressed his wife. The constable agreed that the activity "is just crazy around here" and then arrested him. Later that day, the police were sent to confiscate any guns or ammo at the Mildenstein farm. NAL Energy evacuated its drilling operation and called in sniffer dogs. From the police station, Carsten left an urgent message for Kimberly on their home answering machine.

While the police detained her husband as a suspected terrorist, Kimberly Mildenstein went to a doctor's appointment, attended a son's music festival, and visited with her mother. As soon as she got her husband's message, Mildenstein drove to the Sundre police station, where the RCMP arrested her, fingerprinted her, and charged her with two counts of uttering threats. The arresting officer said he felt sorry for the community coach, but then asked, "Would you want to see a world of chaos?" Mildenstein looked at the man in wonder. She already lived in that dreadful world. She told the constable that "nobody had policed anything" during the fracking boom. A pre-sentencing report would later claim that Mildenstein was "definitely caught up in the environmental movement," though the mother belonged to no political or environmental group of any kind. For weeks after her arrest, Mildenstein says, she felt nothing but shame, embarrassment, and humiliation. She soon became known as the "Crazy Lady in Eagle Hill."

On the evening of the day Mildenstein was arrested, Jessica Ernst was scheduled to give a talk on "The Truth of Hydrofracturing" at Eagle Hill Hall, near the Mildensteins' home. More than 130 farmers and landowners crowded the place. Ernst had heard about the incident, and she began her presentation by speaking out against violence, repeating what she had told citizens in New Brunswick and Ireland. "Don't let fracking put you right back to where your governments want you," warned Ernst. Violence only gave governments an excuse to further threaten and abuse communities, she said. But she also understood the panic of mothers and fathers who worried about groundwater and the health of their children. "The ERCB is only out there for profit," she told the crowd. "We are expendable." The audience gave her a standing ovation.

295

Mildenstein was too fearful to attend the meeting, but she later phoned Ernst. The two women talked for hours about industrial violence and the duties of mothers. Mildenstein cried on the phone, telling Ernst that she had acted out of character by wording her ERCB objection in a crazy way. She shared her deep feelings of shame as well as her fear that her husband might leave her.

In turn, Ernst told Mildenstein about all the other mothers who had called her over the years. Mothers have a sacred duty to protect their children, Ernst said, and it appeared to her that Mildenstein had broken the law to heed that fundamental duty. Had the ERCB done its job and regulated the frackers, Mildenstein would never have filed her threatening objection. In her view, Ernst told the distraught mother, the government had charged Mildenstein in order to terrify and silence other potential critics.

Two months after Mildenstein's arrest, Tracey McCrimmon, the executive director of SPOG, gave an impassioned defense of hydraulic fracturing before Sundre's town council. The council asked her one specific question: how could Alberta avoid the water contamination issues now plaguing fracked communities in Colorado? McCrimmon said that was easy: Alberta had world-class regulators and different hydrocarbon formations. She swore that excellent cementing jobs sealed off the wellbores and that industry neatly controlled all of its fractures. "There has never been a documented case of hydraulic fracturing activities contaminating groundwater in Alberta," declared McCrimmon. She didn't mention Jessica Ernst, the Lauridsens, Debbie Signer, Bruce Jack, Dale Zimmerman, or the Campbells. Nor did she make reference to the hundreds of gag orders or the absence of proper scientific groundwater monitoring in the province. Not one politician challenged her response.

In contrast to Ernst's protracted lawsuit, Mildenstein's case encountered no gatekeepers; it went to court in just seven months. Mildenstein pleaded guilty to uttering a threat, and her lawyer applied for conditional discharge so that the former social worker wouldn't have a criminal record. The Crown opposed a conditional discharge and argued for a criminal conviction as well as a fine. On the day of the sentencing, Jessica Ernst drove on icy roads to Sundre to lend her moral support.

Having considered the facts, the judge ruled that Mildenstein had acted on impulse "arising on a background of stress, anxiety, and frustration at being unable despite significant effort to address genuine public and private safety issues that had arisen in her community." The judge entered no conviction but put Mildenstein on probation for a year. Jessica Ernst thought she had finally entered a court that was concerned with justice. Paddy Munro saw it differently. He concluded that the ERCB had achieved their goal of community intimidation: "They were showing everyone who was in control."

After her ordeal, Kimberley Mildenstein moved her family and took a psychology course to upgrade her educational credentials. Two items discussed in the course particularly caught her attention. One was the issue of bystanders who witnessed bullying but did nothing out of fear. Mildenstein realized that most rural Albertans had unwittingly become part of a "bystanders' culture." Everyone knew someone or had a relative or spouse who worked for the powerful oil and gas industry. As a result, they were afraid to speak out or to challenge bad industry behavior. Although her neighbors agreed that the fracking boom had deeply changed their community, threatened groundwater, and fragmented the land, "they couldn't put their names on anything," Mildenstein says.

The second idea that intrigued her was malpractice. She concluded that groups like SPOG "were falsely representing support, and that's malpractice. They would listen like a bad psychologist but would never offer any support or solutions. They'd invite you back again and repeat the trauma. It's a recipe for suicide." Like Ernst, she concluded that synergy groups amounted to a form of social engineering created to smother dissent.

Midlenstein's brush with the law left an indelible impression. The mother and hockey coach no longer believes she lives in a democracy. Her loss of faith in government and regulators remains an unsettling fracture in her life. Mildenstein attended all three court hearings for Jessica Ernst, and the tone of the proceedings dismayed her. She feels that Ernst has been snared by a legal system with no interest in justice.

Kimberly Mildenstein's views on fracking are unchanged. "The impacts are way too dangerous," she says. "You can't give other people cancer and threaten their water supplies and survival." Whenever she thinks about the events that led to her arrest, she arrives at the same conclusion: "When you are threatened, you are human."

IN 2009, A young anthropologist named Simona Perry began studying the transformation of northeastern Pennsylvania's

Bradford County. Prior to the shale gale, the county had been poor, white, agricultural, patriotic, retiring, and religious. Many people had ties to the land going back to the American Revolution, and many had served in the military. In 2007, industry applied for 99 drilling permits in Bradford County. Then came the frackers, with their trucks, traffic, and temporary workers. By 2010, the number of drilling permits totaled 3,445.

At first, county residents welcomed the industry as an economic savior. People believed industry assurances that natural gas was safe, and they welcomed the money: nearly $160 million in land leases. Most had confidence in the regulators. Many landowners felt it was their patriotic duty to allow industry on their land, to help achieve U.S. energy independence. But a growing number of traffic accidents, skyrocketing rents, and diesel pollution raised doubts, then concerns. The crime rate in the county went up 40 percent. An alarming number of homes sprouted water storage containers, called "buffaloes," and methane venting systems. As the fracking intensified, the number of dairy cows dramatically declined; the industrialization of the land had lowered its carrying capacity.

By 2010, Perry had recorded a significant change in the community. People expressed fears of loss and dread about the future. Everything that had defined the quality of life in the county, from clean water to fresh air, had somehow been eclipsed by the shale gas industry. Farmers complained that what industry promised at the kitchen table rarely matched fracking practices in the field. Neighbor had been pitted against neighbor.

To Perry, the distress experienced by some residents sounded like the trauma of people abused by violent spouses. Landowners told her the pace and scale of fracking had

damaged the basic social tissues of rural life. People's relation-
ships to the land, to their neighbors, and to the community had
all suffered. The industry had stolen something from Bradford,
something tranquil and deep-rooted. One woman told Perry,
"It feels like we are losing our love."

During her research, Perry was struck by how the cycle of
abuse played out in communities fracked by the oil and gas
industry. It happened like clockwork. The cycle began with rising tensions and fears at public meetings and around kitchen
tables as the industry invaded with impossible promises. Once
a frack job inevitably contaminated a well, or a pipeline leaked
into a river, or a truck hit and killed some cows, rage and blame
followed, as did threats and intimidation from industry and its
supporters. During the third part of the cycle, regulators reluctantly got involved. Apologies, denials, and excuses blanketed
the community. Reconciliation was usually accompanied by
the donation of money. A corporate check invariably went to
a hospital, school, library, sports center, or fire department. In
many cases, gag orders were signed in return for cash settlements. Then the community would forget the event until the
cycle repeated itself, snaring more rural victims. It was just as
ERCB lawyer Glenn Solomon had described to a disbelieving
Brent O'Neil.

The cycle left in its wake a conquered people vulnerable to
depression, addiction, and suicide. "There is no doubt in my
mind that abuse on a massive scale is occurring within Bradford County," said Perry at a 2011 summit on the impacts on
fracking. "Until the perpetrators of such abuse are reformed or
brought to justice, there will be no end to the cycle." The U.S.
shale gas industry later characterized Perry's research "as baseless speculation by another NIMBY posing as a scientist."

In 2004, at a meeting at the Rosebud Church, Jessica Ernst had handed out a one-page sheet on the cycle of violence employed by Encana during the unconventional gas boom. The one-pager depicted the same cycle of abuse Perry was to outline. It began with tensions, followed by explosions; it ended with reconciliations, along with gifts from the abuser, such as the $150,000 to the Rosebud Theatre. In her handout, Ernst asked how long a community could let this cycle continue.

Jessica Ernst has not met Simona Perry or visited Bradford County, the second most drilled and fracked county in Pennsylvania. But she considers Perry another sister in an increasingly fractured world. Ernst understands well how that dispossession robs the soul. She says that she hasn't felt at home for years.

Completions

A GAINST ALL ODDS, on November 7, 2014, Jessica Ernst scored a victory in the courts. Her phone rang, and Ernst found both of her lawyers on the other end of the line. "It's Murray here," Klippenstein said, his voice breaking. "Jessica, we won."

In a lengthy decision, Chief Justice Neil Wittmann had granted Ernst the right to sue Alberta Environment for negligence. To emphasize his displeasure at how the government had wasted the plaintiff's and the court's time, Wittmann had also awarded Ernst triple costs. Such an award was rare, Klippensten told his client, though to Ernst that aspect of the decision seemed a merely Pyrrhic gesture. It amounted to $8,500 out of the more than $300,000 she had spent on the lawsuit so far.

Wittmann's ruling found that the government couldn't claim immunity from civil action, because the legislation governing environmental protection in Alberta wasn't as comprehensive as the ERCB's Section 43. The legislation did not, for example, apply to actions taken in bad faith. The idea that a government owed a general duty of care only to the public and none to landowners with contaminated groundwater was also mistaken, said Wittmann. "While this is a novel claim, I find there is a reasonable prospect Ernst will succeed in establishing that Alberta owed her a prima facie duty of care." In a press release, Jessica Ernst called the decision a victory for water and for democracy: "The decision means that landowners can stand up and hold governments and regulators to account if they fail in their duty to properly investigate environmental contamination."

The win catapulted Ernst's case into Canada's legal limelight. An analysis by a University of Calgary law professor said Ernst's lawsuit was shaping up to be "the legal saga of the decade." Borden Ladner Gervais, Canada's largest national full-service law firm, agreed. The firm's brief on the top ten legal decisions affecting the oil and gas industry in 2014 included the Wittmann decision: "At this stage of the proceedings, the Ernst case has brought into focus the potential for regulator or Provincial liability arising out of oil and gas operations," explained the brief. "If Ernst proceeds to trial, it will likely provide more guidance on the scope of the duty of care... required by the Province and the oil and gas operator to discharge their duties in the context of hydraulic fracturing."

After the ruling was announced, Ernst received congratulations from citizens around the world. Some called it a "miracle." In England, community groups celebrated the decision on Facebook: "The wonderful Jessica Ernst—who visited Ireland, Lancashire and Sussex to warn us of the unconventional fossil fuel industry's obscene designs on our communities—has worked exhaustively at this for many years and is a beacon for us all." Much to Ernst's amazement, an editorial in the *Edmonton Journal* congratulated the scientist and called the decision "a victory for the little guy." From central Alberta, Shawn and Ronalie Campbell, who had also lost their water to fracking, sent Ernst their profound thanks: "You stood up for all of us and we are deeply grateful. Celebrate and then go after the next big fish!" 303

Ernst didn't celebrate for long. A week after the Wittmann ruling, her lawyers applied to the Supreme Court of Canada to challenge the Alberta Court of Appeal decision that excluded the ERCB from the lawsuit. To Ernst, the ERCB remained the most guilty party in her lawsuit, and an agency with a closet full of incriminating data on hydraulic fracturing. On April 30, 2015, the Supreme Court agreed to hear her case. The decision both stunned and exhilarated Ernst.

"This case is about whether a government regulator can be held accountable for breaching fundamental and constitutional free speech rights of a landowner," said Cory Wanless to the media. Shortly afterwards, Albertans voted out the corrupt party that had ruled the province for forty-four years.

A MAJOR SHALE gas study in *Science* magazine confirmed Ernst's early observations about the destructive nature of gag orders: "Confidentiality requirements dictated by legal investigations, combined with expedited rates of development and the limited funding for research, are major impediments to peer-reviewed research into environmental impacts." One prominent lawyer estimates that, over the past decade, industry has negotiated hundreds of confidentiality agreements in Alberta alone.

On the issue of groundwater contamination, the oil and gas industry made some unexpected admissions. "Many worries about water quality are based on past operations involving coal-bed methane—shallow deposits in closer proximity to groundwater," said Alex Ferguson, a spokesman for the Canadian Association of Petroleum Producers, in a 2014 *Calgary Herald* advertisement. These operations did occasionally pollute water resources, he added. "In some of the more infamous instances, affected landowners could light their well water on fire."

Nonetheless, Alberta's regulators still swear there hasn't been a single case of water contamination related to fracking. The province's delinquent investigation of groundwater contamination remains the rule, not the exception, across North America. Earthquakes caused by fluid injection increasingly unsettle rural communities. The U.S. Geological Survey now encourages citizens in the central United States to take part in earthquake drills and to learn how to "Drop, Cover and Hold On." In January 2015, high-volume fracking operations in the Duvernay Formation in northern Alberta made global history by creating the largest fracking quake yet, with a magnitude of 4.4. The quake shook homes like cars on a roller-coaster. No one knows what it did to aquifers or other formations.

Around the world, the oil and gas industry continues to thwart local governance and undermine regulations. In Australia, whistleblower Simone Marsh is still seeking justice. The Queensland Senate recently placed a gag order on the spoken evidence and documents she provided on the fraudulent approval process for fracking coal. After heavy lobbying by frackers, the British government introduced a new amendment to its Infrastructure Act in 2015. By changing the trespass law, the amendments erode the eight-hundred-year-old Magna Carta, which proclaimed that no man should be stripped of his rights or possessions. The new act gives industry automatic right of access to "deep-level land" under people's homes without their permission, and grants industry the right to "maximize" hydrocarbon recovery. It also allows companies to "store and leave" products or waste in citizens' backyards. Critics have called the law a perfect example of modern-day fascism.

305

In her home community of Rosebud, Jessica Ernst is no longer shunned. Many residents now support what some call "the scary lawsuit" by donating foodstuffs to Ernst or by quietly appearing at her court hearings. On a global scale, resistance to the spread of hydraulic fracturing has intensified. To date, not one exploratory shale well has been drilled in Ireland. By the end of 2014, as a result of public opposition, the governments of New Brunswick, Nova Scotia, Newfoundland, and the Yukon had imposed restrictions or bans on fracking. New York State outlawed the mining of gas shale basins, to protect public health. It's noteworthy that Jessica Ernst gave talks in all of these jurisdictions.

After France banned hydraulic fracturing in 2011, it studied the mining of coal beds for gas. Two French government agencies later found that the unconventional resource presented significant risks to groundwater and public health even without hydraulic fracturing: gases could migrate to the surface and cause explosions and suffocate or poison citizens.

In Australia, the coal-fracking industry has created the largest and most diverse protest movement in that nation's history. "Lock the Gate" urges landowners to deny access to coal-seam gas mining companies and to advocate for a moratorium until rigorous research has been done on the industry's impacts on water, land, and people's health. As a 2013 review of coal-seam gas activities in New South Wales noted, "There is a widespread perception that industry and government are, at worst, colluding against the public's best interests." John Williams, one of the country's top scientists, has asked, "Do we want degraded and collapsing landscapes? If the answer is yes, then we appear to be well on the way."

Gas migration from aging oil and gas infrastructure in the Los Angeles Basin remains a chronic problem, accounting for 8 percent of all methane emissions in the city. California's Environmental Protection Agency has identified 2,500 wastewater disposal wells that are injecting fracking and other waste into protected aquifers. Douglas Hamilton, who studied the causes of the Fairfax explosion, says authorities at the time willfully ignored the findings that man-made fractures connect to existing ones: "The industry didn't say anything; nor did the regulator." It's not a story that the economic powers were "interested in seeing pursued," he explains.

Fracking operations continue to pollute groundwater. At a meeting of the American Chemical Society in August 2014, two Stanford researchers described how industry injected "thousands of gallons of undiluted diesel fuel and millions of gallons of fluids" containing salt water, methanol, and solvents into underground sources of drinking water during a tight gas mining in Pavillion, Wyoming.

In June 2015 the U.S. Environmental Protection Agency finally released a draft of its long awaited second report on fracking. The agency's findings, which largely omitted data from Pavillion, Wyoming, completely reversed the conclusion of its fraudulent 2004 report. This time the EPA confirmed, as it did in an earlier 1987 report, that the fracturing of unconventional wells had "led to impacts on drinking water resources including contamination of drinking water wells."

In just one documented case in northeastern Pennsylvania the agency found that 25 percent of 36 water wells had been contaminated by methane due to industry activity. The report also admitted that industry has knowingly fracked into aquifers containing drinking water. It even cited Jessica Ernst's case: "In one field in Alberta, Canada, there is evidence that fracturing in the same formation as a drinking water resource...led to gas migration into water wells," said the report. Despite having limited data on quality of groundwater prior to fracking and "the inaccessibility of some hydrofracturing activities" due to confidentiality agreements, the report inexplicably concluded that contamination was not "widespread."

In a recent presentation Usman Ahmed, the vice president of Baker Hughes, a major fracking service company, described the technology as a hit and miss affair. Although the industry had increased the number of stages from 15 to 20 and had extended the length of lateral wells from 2000 to 4500 feet, unconventional wells still experienced decline rates of 80 percent. Furthermore Ahmed admitted that 70 percent of unconventional wells do not reach their production targets; that 60 percent of all fracture stages are ineffective and that 73 percent of all drillers don't know enough about existing faults and fractures in the subsurface.

Encana's "resource play" strategy has proven an unqualified failure. As the massive fracking of shale rock drove natural gas prices down from $14 to $2 a gigajoule, the company struggled with high debt loads. In 2013, CEO and president Randy Eresman abruptly resigned, and a U.S. Securities and Exchange Commission filing disclosed that Encana had invested more in shale plays "than can be optimally developed." As a consequence, the company sold off billions of dollars' worth of assets and laid off 20 percent of its staff. To avoid financial ruin, the firm switched from fracking rock for dry gas to pulverizing formations for oil and condensates. In 2014, after being charged with violating Michigan's antitrust laws over rigged real estate deals related to fracking, the company paid $5 million in fines to avoid more charges.

Since the Industrial Revolution, an event fueled by the mining and fracking of coal seams, methane levels have been rising in the atmosphere. As the natural gas industry expanded in the 1970s, methane levels jumped from 830 to 1,500 parts per billion. The dewatering and fracking of the San Juan Basin has made it more permeable by 10 to 100 times; NASA calls it the largest "methane anomaly" on the continent. With the cracking of shale basins in the United States and the mining of coal basins throughout China, methane levels have now reached a sobering 1,800 ppb, the highest level in 650,000 years. To date, not one energy regulator has assessed the impact of fracking on global methane emissions and climate change. There appears to be no duty of care.

Anthony Ingraffea now denounces fracking as an extreme and dangerous technology: "The industry is reaching into the deepest, darkest corner of its almost empty world hydrocarbon warehouse, and using an inelegant, inefficient, wasteful, bludgeoning process to keep itself alive, at the expense of exacerbating climate change." He places Jessica Ernst's lawsuit on a special pedestal because it has exposed the industry's Achilles heel: migrating gases from millions of leaky wells.

From the outset, industry denials have followed a similar pattern, Ingraffea says. "One: We didn't do it. Two: Prove it.

Three: Silence. Maybe we did it, but we're not going to tell anybody." Ernst's case and her refusal to settle out of court have exposed the deception: "Number one, she caught the industry doing it. Number two, she's got the evidence to prove it, and number three, she won't be silent." For all of this, Ingraffea regards Ernst as a public hero.

The evolution of fracking technology would not have surprised Jacques Ellul. In his final book on the subject, *The Technological Bluff*, the Christian philosopher argued that the rapid adoption of techniques, from computers to genetic engineering, generates totalitarian discourses: "The more indispensable [these techniques] become, the more power they have, the more important they are, the more money they make, the more difficult they are to uproot. Their propagation becomes... an expression of both their self-interest and the strengthening of their situation. They cannot act in any other way. They are forced to reject increasingly what remains of democracy." The constant proliferation of techniques, argued Ellul, makes a society more complex, more wasteful, and more disorderly. For Ellul, only two forms of resistance mattered: "We must be prepared to reveal the fracture lines and to discover that everything depends on the qualities of individuals."

TO MANY, HYDRAULIC fracturing is additional evidence that humans have become a radical force in altering the geology of the planet. Scientists call this new epoch the Anthropocene. Fossil fuel emissions have destabilized the climate and changed the carbon cycle. Engineers have commandeered 50 percent of the world's available fresh water through dams and other diversions. Humans have used cheap energy to industrialize half of the planet's surface area. Plastics choke the oceans, and crushed aluminum soda cans and concrete make their own geological strata.

Fracking has extended the reach of the Anthropocene to the earth's crust. Engineers can now crack shallow geologies the size of European countries or blast shale rock two miles below the earth's surface. To support the process, the industry has injected lakes of water and chemicals underground (250 billion gallons since 2005) and has pumped another 280 billion gallons of salty wastewater to the surface. The massive injections have geo-engineered "aseismic" Oklahoma into a landscape more earthquake-prone than California.

Forty years ago, the oil and gas industry created a fiery hole in Turkmenistan, in the middle of the Karakum Desert. Locals still call it the Door to Hell. The mysterious crater in the Derweze Oil Field, a bucket-list destination for aging tourists, resembles an open furnace whose fiery heat dominates the skyline at night. The spectacular flaming chasm is 226 feet wide and 98 feet deep. The stories about its origins vary, but most agree that a Soviet drilling rig disappeared into a sinkhole on the site around 1971. After the mishap, gases vented from the sinkhole so prolifically that engineers decided to set the whole thing on fire. They expected the flaring to last a few days, but the methane continues to burn like some kind of eternal flame. The evidence suggests that industry accidently drilled into a natural fracture that was carrying methane from the breathing earth.

Some people say that the Soviet drilling rig can still be found on the other side of the Door to Hell. Jessica Ernst thinks that Canada's legal system and the global fracking industry are there with it. But no matter how hot things may get at the Door to Hell, she says, "I will not stand down."

AUTHOR'S NOTES

I FIRST MET Jessica Ernst in 2004 while reporting for the *Globe and Mail's Report on Business Magazine*. At the time I warned her, as both a landowner and a journalist, that the unprecedented industrial carpet bombing unfolding in her backyard might change the course of her life forever. I advised her to leave Rosebud, Alberta. Instead she stayed. And then she took a stand against her former client and the world's most powerful industry.

Since then our lives have often crossed. As a concerned landowner, I helped organize several public meetings on the fracking of unconventional rock between 2004 and 2006. Ernst spoke at many of these events. And over the years we often corresponded and shared our research findings on methane contamination of groundwater and gas migration.

I had no intention of writing this book until it became apparent that the scale of industrial drilling and fracking that I, Jessica Ernst, and many other rural Albertans witnessed in 2004 was but a prelude to the dramatic shale gas revolution still unsettling North America today. At that point Ernst's

ordeal became a troubling and important window on a bru-
tal North American drama. The tortured path of her lawsuit
also suggested that only a book, not another magazine article,
might offer some measure of real justice.

Jessica Ernst shared all of her voluminous documentation,
including emails, tape recordings, and freedom of information
requests. Moreover, she answered all questions with a fearless
and critical honesty. As a consequence the book reflects her
point of view and portrays her sacrifice. It is a profile of one
unique woman's courage.

Fifteen percent of all book royalties will be set aside to help
fund her ongoing legal case.

Parts of the book have drawn upon the works of two
Alberta journalists: Jeremy Klazsus and Tadzio Richards. Klazsus wrote about Debbie Signer's dismal adventures with
Alberta Environment and graciously shared his transcripts and
notes. Together with Cameron Esler, Tadzio Richards made a
documentary about Fiona and Peter Lauridsen, neighbors of
Jessica Ernst. Their experience with methane-contaminated
water directly mirrors that of Jessica Ernst. The matter-of-fact
documentary, Burning Water, captures the nuances and dysfunc-
tion of a true petrostate.

A number of key players in this story were interviewed
directly by the author over the period of a decade. They include
Douglas Hamilton, George V. Chilingar, Jonathan Wright, Mat-
thew Biren, Stacy Knull, Mark Taylor, Laurel Lyon, Ed McCord,
Cynthia McMillian, David Ludder, Fiona and Peter Lauridsen,
Ronalie and Shawn Campbell, David Swann, Anthony Ingraf-
fea, Helen Rezanowich, Karlis Muehlenbachs, Brent O'Neil,
Ann Craft, John Cherry, Carl Weston, Dr. John O'Connor,
Maurice Dusseault, Bruce Jack, Diana Daunheimer, Kimberly

Mildenstein, Laura Amos, Tweeti Blancett, Gwen Lachelt, and Oscar Steiner. Many working members of the oil patch also shared important information.

The critical work of two geologists, David Hughes and Art Berman, shape this book. Both experts recognize unconventional resources as a signature of peak oil and a major descent down the energy pyramid. And both men have carefully documented how fracking shale or coal requires a costly drilling treadmill that delivers ever diminishing energy returns overtime. Berman, in particular, views "the shale gale" as a temporary phenomenon that will amount to nothing more than a "retirement party" for the industry.

Ernst's lawyers, Murray Klippenstein and Cory Wanless, generously sat through interviews and quickly answered emails. I watched them patiently construct a remarkable lawsuit with a demanding and meticulous client. Both Murray and Cory belong to a rare breed of professionals who still think that the public interest matters in an age of corporate feudalism. Although Klippenstein thought some parts of the book did not adequately explain legal complexities or fairly represent the spirit of his firm, I respectfully disagree. I simply tried to avoid legal jargon and chose to express some technical ideas in common language. The book also reflects the point of view of its central subject: the indomitable Jessica Ernst. She greatly admires her lawyers but has little regard for Canada's legal system.

Last but not least, I owe many thanks to several people who helped row this difficult book to shore. Rob Sanders, Greystone's publisher, kept the funds flowing even when several missed deadlines invited a touch of apoplexy. The Canada Council also provided an essential writing grant, which kept my oars in the water. Adèle Hurley, director of the Program on Water Studies at the University of Toronto's Munk School of Global Affairs, was as supportive as ever. A good chunk of the book was started while I served as the writer in residence at the Haig-Brown House in Campbell River, British Columbia, in the winter of 2014. There are few places as magical and regenerative. The residency, beautifully operated by the Museum at Campbell River, unexpectedly changed our lives.

Throughout this project, Dave Beers, my friend and the fearless founder of The Tyee, kept me gainfully employed at Canada's best independent paper. At The Tyee, Robyn Smith, an excellent editor, handled many unfolding legal and fracking stories about the Ernst lawsuit with aplomb. Barbara Pulling, with whom I have now written four books, executed another brilliant edit for Greystone Books. (Bravo, Barbara.) John Readwin, my long-time office buddy and a digital genius, kept my ancient MacBook Pro working and often lifted my weary spirits with tales of good living and grand feasts. Michel Clario offered some brilliant ideas and comic insights. "Annie57" provided critical feedback on the final manuscript and was an excellent reader over my shoulder. Stephanie Fysh did another fine copyedit. Ben Parfitt offered encouragement while being lovingly present for the late Alicia Priest, an inspiration to all of her friends. Primrose and Asher, my beloved Anatolians, walked me home each night.

Doreen, my profile in courage and a remarkably direct Scot, endured, lived, and loved through all the rowing. The master navigator not only picked the right title for this book but kept it on course. She made all of this possible, and more. No voyage is complete without her.

Readers can contact the author at andrew@andrewnikiforuk.com.

A NOTE ON
NAMES AND LANGUAGE

MANY OF THE key government agencies that Jessica Ernst has challenged have gone through several Orwellian name changes.

Ernst's historical water well records were filed with Alberta Environmental Protection. It then morphed into Alberta Environment and, later, Alberta Environment and Water. It is now called Alberta Environment and Sustainable Resource Development. Throughout the book it is simply referred to as Alberta Environment.

The scandal-plagued energy regulator has undergone similar brand-name surgery. At the beginning of this story it was called the Energy Utilities Board. After it was caught spying on Albertans, it became the Energy Resources Conservation Board. It now calls itself the Alberta Energy Regulator and is 100 percent funded by industry. Throughout the book the board is called what it was called at the time.

Fracking is not a real word, but the world and industry's aggressive tactics have made it so. Industry slang for the forceful technology used to be *frac'ing* or *fraccing*. But when the technology got bigger and messier with slick-water fracturing, the media adopted the more familiar-looking *fracking*.

Fracking, of course, was a common curse word on the television series *Battlestar Galactica*. As a consequence, many petroleum engineers still refuse to use it. Yet whether *fracking* or *frac'ing*, the term still means the same thing: the injection of fluid or gas through vertical or horizontal wells into shallow or deep formations to pulverize low-quality rocks.

To this day, many environmental groups wrongly assume that the word only refers to cracking deep rock formations with horizontal wells. Meanwhile, industry continues to sexualize the technology by referring to it as *stimulation* or *enhanced stimulation*. A "nitrogen stimulation," for example, is a hydraulic fracturing job with gas. Rig hands often directly call the technology "earth fucking." It is an apt description.

In any case, the book mostly refers to the technology as "hydraulic fracturing." And where it uses the new media creation *fracking*, petroleum engineers can relax: it is not being employed as an expletive.

CRITICAL SOURCES

CHAPTER ONE: The Dress for Less Explosion

Chazanov, Mathis. 1990. Explosive gas still imperils Fairfax area, experts say. *Los Angeles Times*, January 7.

City of Los Angeles, 1989 Methane Gas Task Force. 1990. *Task Force Report II on the Methane Gas Incursion, Fairfax Area, City of Los Angeles.* City of Los Angeles.

City of Los Angeles, Task Force on the March 24, 1985 Methane Gas Explosion and Fire in the Fairfax Area. 1985. *Task Force Report on the March 24, 1985 Methane Gas Explosion and Fire in Fairfax Area.* City of Los Angeles.

Endres, Bernard, George V. Chilingar, and T.F. Yen. 1991. Environmental hazards of urban oilfield operations. *Journal of Petroleum Science and Engineering* 6: 95–106. doi:10.1016/0920-4105(91)90030-Q.

Gurevich, A. E., B. L. Endres, J. O. Robertson, Jr., and G. V. Chilingar. 1993. Gas migration from oil and gas fields and associated hazards. *Journal of Petroleum Science and Engineering* 9: 223–28. doi:10.1016/0920-4105(93)90016-8.

Hamilton, Douglas H., and Richard L. Meehan. 1971. Ground rupture in the Baldwin Hills. *Science* 172, no. 3981: 333–44. doi:10.1126/science.172.3981.333.

Hamilton, Douglas H., and Richard L. Meehan. 1992. Cause of the 1985 Ross Store explosion and other gas ventings, Fairfax District, Los Angeles. *Engineering Geology Practice in Southern California*. Special Publication No. 4. Belmont, CA: Association of Engineering Geologists, Southern California Section.

Khilyuk, Leonid F., John O. Robertson, Jr., Bernard Endres, and G. V. Chilingar. 2000. *Gas Migration: Events Preceding Earthquakes*. Houston, TX: Gulf Publishing.

Murphy, Dean. 1989. L.A. agrees to pay 20 hurt in Fairfax explosion. *Los Angeles Times*, July 20.

Perera, Dave. 2001. Fresh produce and streets of fire: Making sense of the methane explosion in Fairfax. *LA Weekly*, May 2.

Ramos, George, and Stephen Braun. 1989. Major methane gas leak closes shopping strip. *Los Angeles Times*, February 8.

Ramos, George, and Steve Harvey. 1985. Gas explosion shatters Fairfax store; 23 hurt. *Los Angeles Times*, March 25.

Schoell, Martin. 1983. Genetic characterization of natural gases. *AAPG Bulletin* 67: 2225–38.

Schoell, Martin, P. D. Jenden, M. A. Beeunas, and D. D. Coleman. 1993. Isotope analysis of gases in gas field and gas storage operations. *Proceedings of the Society of Petroleum Engineers Gas Technology Symposium, Calgary, Alberta, Canada, June 28–30, 1993*, pp. 337–44. SPE No. 26171. doi:10.2118/26171-MS.

CHAPTER TWO: This Much You Should Know

Allwright, Tony. 2012. Tallrite debates fraccing on TV. Tallrite Blog, March 13. At www.tallrite.com/weblog/archives/Q1-2012.htm.

Hymas, Kay. 1983. Akokiniskway: "By the River of Many Roses …" Rosebud, AB: Rosebud Historical Society. At www.ourroots.ca/e/toc.aspx?id=4375.

Jaremko, Deborah. 2005. Fur-bearers in the field: Study examines cumulative effects of activity at Chinchaga. *Oilweek*, August.

Mischel, Walter, and Ozlem Ayduk. 2004. Willpower in a cognitive affective processing system: The dynamics of delay of gratification. In *Handbook of Self-Regulation: Research, Theory, and Application*, edited by Roy F. Baumeister and Kathleen D. Vohs. New York: Guilford Press.

Mischel, Walter, Yuichi Shoda, and Monica L. Rodriguez. 1989. Delay of gratification in children. *Science* 244, no. 4907: 933–38. doi:10.1126/science.2658056.

318

CHAPTER THREE: Fracking Oildorado

American Oil & Gas Historical Society. n.d. Shooters—A "fracking" history. At aoghs.org/technology/hydraulic-fracturing/.

Armstrong, Kevin, Roger Card, Reinaldo Navarrete, et al. 1995. Advanced fracturing fluids improve well economics. Oilfield Review 7, no. 3: 34–51.

Black, Brian. 2000. Petrolia: Landscape of America's First Oil Boom. Baltimore, MD: Johns Hopkins University Press.

CAEPLA. 2012. A Revolution Underground: The History, Economics & Environmental Impacts of Hydraulic Fracturing. Regina, SK: Canadian Association of Energy and Pipeline Landowners Associations.

Economides, Michael J. 2011. Hydraulic fracturing: The state of the art (and the market and the technology and the environment). Energy Tribune, August 26. At www.energytribune.com/8672/hydraulic-fracturing-the-state-of-the-art-2.

Howard, G. C., and C. R. Fast, eds. 1970. Hydraulic Fracturing. Monograph Vol. 2, Henry L. Doherty Series. New York: Society of Petroleum Engineers.

Hubbert, M. King, and David G. Willis. 1957. Mechanics of hydraulic fracturing. Transactions of the Society of Petroleum Engineers of AIME 210: 153–68.

King, George E. 2012. Hydraulic fracturing 101: What every representative, environmentalist, regulator, reporter, investor, university researcher, neighbor and engineer should know about estimating frac risk and improving frac performance in unconventional gas and oil wells. Paper presented to the SPE Hydraulic Fracturing Technology Conference, February 6–8, The Woodlands, TX. SPE 152596. Society of Petroleum Engineers.

Mader, Detlef. 1989. Hydraulic Proppant Fracturing and Gravel Packing. New York: Elsevier.

McLaurin, John J. 1902. Sketches in Crude-Oil: Some Accidents and Incidents of the Petroleum Development in All Parts of the World. Franklin, PA.

Montgomery, Carl T., and Michael B. Smith. 2010. Hydraulic fracturing: History of an enduring technology. Journal of Petroleum Technology (Society of Petroleum Engineers) 62, no. 12: 26–32.

Nordyke, Milo D. 1998. The Soviet program for peaceful uses of nuclear explosions. Science & Global Security 7, no. 1: 1–117.

Pennsylvania Heritage Magazine. 2009. The shooting stars of Drake Well. Vol. 35, no. 3.

Tallini Tales of Destruction. 2000–2010. Otto Cupler Torpedo Co.; AnaLog Services, Inc. At www.logwell.com/tales/menu/.

CHAPTER FOUR: Before Shale: Coal

Atkinson, W. N., and J. B. Atkinson. 1886. *Explosions in Coal Mines.* London: Longmans Green.

Associated Press. 1993. Company refuses to shut down wells: Methane leak detected. *Daily Times* (Farmington, NM), November 12.

Ayoub, Joseph, Jerry Hinkel, David Johnston, and Jeffrey Levine. 1991. Learning to produce coalbed methane. *Oilfield Review*, January: 27–40.

Bland, Douglas. 1992. Coalbed methane from the Fruitland Formation, San Juan Basin, New Mexico. In *San Juan Basin* IV, edited by S. G. Lucas, B. S. Kues, T. E. Williamson, and A. P. Hunt, pp. 373–83. 43rd Annual Fall Field Conference Guidebook. Socorro, NM: New Mexico Geological Society.

Bowen, Kevin. 2001. Burn victim wants to stop gas wells. *Tuscaloosa News,* March 29.

Chafin, Daniel T. 1994. *Sources and Migration Pathways of Natural Gas in Near-Surface Ground Water beneath the Animas River Valley, Colorado and New Mexico.* Water-Resources Investigations Report 94-4006. Denver, CO: U.S. Geological Survey, in cooperation with the Colorado Oil and Gas Conservation Commission, La Plata County, and the Southern Ute Tribal Council.

Colorado Oil and Gas Conservation Commission. 2000. *Summary Report of Bradenhead Testing, Gas Well Remediation, and Ground Water Investigations, San Juan Basin, La Plata County, Colorado.* Colorado Oil and Gas Conservation Commission, May 26.

Davy, Humphrey. 1816. On the fire-damp of coal mines, and on methods of lighting the mines so as to prevent its explosion. *Philosophical Transactions of the Royal Society of London* 106: 1. doi:10.1098/rstl.1816.0001.

Ebner, Dave. 2005. EnCana's "field of dreams" has Colorado locals crying the blues. *Globe and Mail,* July 15.

Elder, Curtis H. 1974. *Degasification of the Mary Lee Coalbed Near Oak Grove, Jefferson County, Ala., by Vertical Borehole in Advance of Mining.* Report of Investigations 7968. Washington, DC: U.S. Department of the Interior, Bureau of Mines.

Elder, Curtis H. 1977. *Effects of Hydraulic Stimulation on Coalbeds and Associated Strata*. Report of Investigations 8260. Washington, DC: U.S. Department of the Interior, Bureau of Mines.

Eltschlager, Kenneth K., Jay W.Hawkins, William C. Ehler, and Fred Baldassare. 2001. *Technical Measures for the Investigation and Mitigation of Fugitive Methane Hazards in Areas of Coal Mining*. Pittsburgh, PA: Office of Surface Mining Reclamation and Enforcement Appalachian Regional Coordinating Center.

Flores, Romeo. 2013. *Coal and Coalbed Gas: Fueling the Future*. Boston: Elsevier Science.

Freese, Barbara. 2003. *Coal: A Human History*. New York: Perseus.

Greene, Susan. 2001. Coal-bed methane fueling dispute. *Denver Post*, September 9.

Halliburton. 2009. *Coalbed Methane Development—A Vital Part of the Total Energy Mix*. Halliburton.

Hamburger,Tom, and Alan C. Miller. 2004. Halliburton's interests assisted by White House. *Los Angeles Times*, October 14.

Hinchman, Steve. 1993. Methane creates an explosive situation in Colorado. *High Country News*, December 27.

Kuuskraa, Vello A., and Hugh D. Guthrie. 2002. Translating lessons learned from unconventional natural gas R&D to geologic sequestration technology. *Journal of Energy & Environmental Research* 2, no. 1: 75–86.

Lathem, Dennis. 2009. LEAF v. EPA: A Challenge to Hydraulic Fracturing Of Coalbed Methane Wells in Alabama. Coalbed Methane Association of Alabama.

Ludder, David. 1999. *A Decade of Efforts to Protect Alabama's Underground Sources of Drinking Water from Contamination by the Methane Gas Industry*. Published by the author.

McGuire, Kim. 2007. "No one is neutral" in water fight. *Denver Post*, August 12.

Methane Recovery from Coalbeds Project. 1981. *Resource Engineering Workshop Minutes, Denver, Colorado, January 21–22, 1981*. Prepared for the United States Department of Energy, Morgantown, WV.

U.S. Bureau of Land Management. 1999. *Coalbed Methane Development in the Northern San Juan Basin of Colorado: A Brief History and Environmental Observations*. U.S. Bureau of Land Management, San Juan Office.

U.S. Department of the Interior. 1975. *Energy Research Program of the U.S. Department of the Interior. FY 1976*. Prepared by the Office of Research and Development, January. Washington, DC: U.S. Department of the Interior.

Wright, Ken. 1993. Fouled water leads to court. *High Country News*, April 19.

CHAPTER FIVE: A Revolution Under Rosebud

Alberta Department of the Environment, Multi-Stakeholder Advisory Committee, Water Working Group. 2004. *Alberta Environment Guidelines for Groundwater Diversion for Coalbed Methane/Natural Gas in Coal Development*. Calgary, AB: Alberta Department of the Environment.

Alberta Energy and Utilities Board. 2004. *Shallow Fracturing Incidents*. Unpublished report.

Alberta Geological Survey. 1990. *Coal Bed Methane in Alberta—What's It All About?* Seminar Proceedings, May 1–2, Westward Inn, Calgary. Information Series No. 108. Alberta Research Council.

Amey, Allan. 2003. Sparking a less carbon-intensive future—Greenhouse gas technologies: Enhanced coalbed methane and biomass power. C3 *Views: Climate Change Central Newsletter*, no. 5: 1–12.

Amos, Laura, and Larry Amos. 2005. *Living in the Gas Fields: One Family's Story*. Presented to the 2005 People's Oil and Gas Summit: Toxics in Our Communities, Farmington, NM. At www.earthworksaction.org/library/detail/living_in_the_gas_fields_one_familys_story.

Asfeldt, Hans. 2012. The Campbells: Alberta ranchers: Methane migration into drinking water aquifer. *Alberta Voices*, October 1. At albertavoices.ca/stories/thecampbells/.

Colburn, Theo. 2002. *An Analysis of Possible Increases in Exposure to Toxic Chemicals in Delta County, Colorado Water Resources as a Result of Gunnison Energy's Proposed Coal Bed Methane Extraction Activity*. Letter to Allen Belt and Robert Storch, October 22. At www.journeyoftheforsaken.com/colborncomments2009.htm.

Curran, B. 2006. *The EUB View: Busting the Myths behind CBM*. Alberta Energy and Utilities Board.

D'Aliesio, Renata. 2006. Troubled waters. *Calgary Herald*, November 18.

Farmer, David, and Gavin Fitch. 2007. Coalbed methane development: Legal and regulatory issues. *Environment Law* 10, no. 2: 1–14.

Kelly, Walton Ross, Gerald Matisoff, and J. Berton Fisher. 1985. The effects of a gas well blow out on groundwater chemistry. *Environmental Geology and Water Science* 7, no. 4: 205–13. doi:10.1007/BF02509921.

Klaszus, Jeremy. 2006. Trouble in the fields. *Alberta Views*, October: 28–33.

Hanel, Joe. 2005. COGCC seeks aid in dealing with wells. *Durango Herald*, December 8.

Happy Valley Surface Rights Association. 2006. Letter to J. Richard McKee, Alberta Energy and Utilities Board, February 18.

Harrison, Samuel S. 1983. Evaluating system for ground-water contamination hazards due to gas-well drilling on the glaciated Appalachian Plateau. *Ground Water* 21, no. 6: 689–700. doi:10.1111/j.1745-6584.1983.tb01940.x.

Harrison, Samuel S. 1985. Contamination of aquifers by overpressuring the annulus of oil and gas wells. *Ground Water* 23, no. 3: 317–24. doi:10.1111/j.1745-6584.1985.tb00775.x.

Henton, Darcy. 2007. Activist "banished" by the EUB. *Edmonton Journal*, July 29.

Hydrogeological Consultants Ltd. 2005. *EnCana Corporation Redland Area NE 10-027-22 W4M Sean Kenny Site Investigation*. File 04-510, January.

Jeffords, Jim. 2005. Testimony, May 19. *Congressional Record—Senate* 151, part 8: 10487–88.

MacLeod, Andrew. 2006. Burning waters: UVic partner's environmental record questioned. *Monday Magazine*, June 21.

McElheran, Graeme. 2005. CPAWS won't wait for government to develop coalbed methane regulations. *Yukon News*, December 2.

Mesley, Nicolas. 2007. En Alberta l'eau s'enflamme! *Le Coopérateur agricole*.

Nijhuis, Michelle. 2006. How Halliburton technology is wrecking the Rockies. *OnEarth*, Summer. At www.nrdc.org/onearth/06sum/rockies1.asp.

Reid, Jim. 2004. *Shallow Operations Issues*. PowerPoint. Presented to the PTAC Shallow Gas Production Technology Forum, May 12. Alberta Energy and Utilities Board.

Rontogiannis, Nick. 2004. Coal bed methane in Western Canada: The coming of age. TD Newcrest, November 17.

U.S. Environmental Protection Agency, Office of Water. 2004. *Evaluation of Impacts to Underground Sources of Drinking Water by Hydraulic Fracturing of Coalbed Methane Reservoirs*. EPA 816-R-04-003, June. Washington, DC: U.S. Environmental Protection Agency.

CHAPTER SIX: Criminal Threats

Alberta Energy and Utilities Board. 2006. *Directive 027: Shallow Fracturing Operations—Interim Controls, Restricted Operations and Technical Review.* January 31.

Calgary, AB: Alberta Energy and Utilities Board.

Bennett, Les, Joël Le Calvez, David R. Sarver, et al. 2005/2006. The source for hydraulic fracture characterization. *Oilfield Review* 17, no. 4: 42–57.

Bonnett, Art, and Demos Pafitis. 1996. Getting to the root of gas migration. *Oilfield Review* 8, no. 1.

Nikiforuk, Andrew. 2005. Life inside a science project. *Report on Business Magazine*, April 29.

Richards, Tadzio. 2007. Burning water. *Maisonneuve*, no. 23.

CHAPTER SEVEN: Banished

Baltiou, Leanard V., Brent K. Warren, and Thanos A. Natros. 2008. State-of-the-art in coalbed methane drilling fluids. SPE *Drilling & Completion*, 23, no. 3. SPE-101231-PA. Society of Petroleum Engineers. doi:10.2118/101231-PA.

Besler, Monte R., John Warren Steele, Tanner Egan, and Jed Wagner. 2007. Improving well productivity and profitability in the Bakken—A summary of our experiences drilling, stimulating, and operating horizontal wells. Presented to the SPE Annual Technical Conference and Exhibition, 11–14 November, Anaheim, CA. SPE-110679-MS. Society of Petroleum Engineers. doi:10.2118/110679-MS.

Fekete, Jason. 2007. Water grief brings cowboy to tears: Rancher fights contamination. *Calgary Herald*, May 2.

Frank, Charles. 2007. EUB woes tip of iceberg. *Calgary Herald*, September 15.

Nikiforuk, Andrew. 2006. Fire water. *Canadian Business*, August 14.

Nikiforuk, Andrew. 2007. Not in our backyard. *Canadian Business*, October 22.

CHAPTER EIGHT: Keys to the Bank

Alberta Hansard. 2006. Coal-bed methane drilling. May 17. Calgary, AB: Legislative Assembly of Alberta.

Gelinas, Grant. 2006. [Bruce Jack contaminated water well explosion]. CBC *News*, Part 1, October 25; Part 2, October 26.

Omni-McCann Consultants Ltd. 2007. *Groundwater Supply Concerns Regarding CBM Development, Wheatland County, Alberta.* Project No. 5-231-1. Wheatland Surface Rights Action Group.

CHAPTER NINE: Fingerprints and Liabilities

Bachu, Stefan. 2007. Carbon dioxide storage capacity in uneconomic coal beds in Alberta, Canada: Methodology, potential and site identification. *International Journal of Greenhouse Gas Control* 1, no. 3: 374–85. doi:10.1016/S1750-5836(07)00070-9.

Blyth, Alexander. 2007a. *Ernst Water Well Complaint Review.* December 31. Alberta Research Council for Alberta Environment.

Blyth, Alexander. 2007b. *Lauridsen Water Well Complaint Review.* December 20. Alberta Research Council for Alberta Environment.

Blyth, Alexander. 2007c. *Signer Water Well Complaint Review.* December 31. Alberta Research Council for Alberta Environment.

Blyth, Alexander. 2007d. *Zimmerman Water Well Complaint Review.* November 16. Alberta Research Council for Alberta Environment.

Budwill, Karen. 2006. Role of biogenic gas generation for sustainable CBM production. Presented to the Williston Basin Petroleum Conference, May 7–9, Minot, ND.

Canada, Standing Committee on Environment and Sustainable Development. 2007. Evidence: Tuesday, May 8. At www.parl.gc.ca/HousePublications/Publication.aspx?DocId=2920058&Language=E&Mode=1.

Gordon, Sue, and Alexander Blyth. 2008. *Campbell Water Well Complaint Review.* January 16. Alberta Research Council for Alberta Environment.

Harris, Michael. 2014. *Party of One: Stephen Harper and Canada's Radical Makeover.* Toronto: Penguin.

McDougall, John. 2007. Briefing—Alberta Research Council's (ARC) involvement with Alberta Environment regarding effects of coal-bed methane activities on groundwater wells in southern Alberta. Memorandum to Hon. Doug Horner, May 3. Alberta Research Council.

Muehlenbachs, Karlis. 2011. Identifying the sources of fugitive methane associated with shale gas development. Presented to Resources for the Future, November 14, Washington, DC. Updated with new data, January 2012. At

www.rff.org/Documents/Events/Seminars/111114_Managing_the_Risks_
of_Shale_Gas/Mulenbachs%20Nov%2014FINAL.pdf.

Nikiforuk, Andrew. 2011. Fracking contamination "will get worse": Alberta
expert. *The Tyee*, December 19.

Palmer, Ian D. 2008. Coalbed methane wells are cheap, but permeability can
be expensive! *Energy Tribune*, March 19.

Saskatchewan Research Council. 1996. *Migration of Methane into Groundwater
from Leaking Production Wells Near Lloydminster. Report for Phase 2* (1995). CAPP
Pub. #1996-0003. Canadian Association of Petroleum Producers.

Schmitz, Ron, Patrick B. Carlson, Garry D. Lorenz, Michael D. Watson, and
Brian P. Erno. 1993. Husky Oil's gas migration research effort: An update.
Presented to the Heavy Oil and Oil Sands Technical Symposium, March
9, Calgary, AB.

Struzik, Ed. 2006. Fighting for the prairie grassland: EnCana is embroiled in
a battle with the military for the right to drill in the Suffield wildlife area.
Edmonton Journal, February 5.

Tilley, Barb, and Karlis Muehlenbachs. 2007. Recognizing natural gas contamination of water wells in a petroliferous region. In *23rd International Meeting
on Organic Chemistry Book of Abstracts, September 9th–14th.* P331-WE.

Tilley, Barb, and Karlis Muehlenbachs. 2008. Letter to MLA David Swann, January 23.

Urbina, Ian. 2011. Drilling down: A tainted water well, and concern there may
be more. *The New York Times*, August 3.

Watson, Theresa, and Stefan Bachu. 2007. Evaluation of the potential for gas
and CO_2 leakage along wellbores. SPE Paper 106817. Society of Petroleum
Engineers. At shale.palwv.org/wp-content/uploads/2014/02/SPE-Pa-
per-106817-Revised-for-publication2.pdf.

Watson, Theresa, and Stefan Bachu. 2008. Wellbore leakage potential in CO_2
storage or EOR. Presented to the Fourth Wellbore Integrity Meeting,
March 19, Paris, France.

CHAPTER TEN: The Police Come Calling

Arsenault, Christopher. 2011. *Loud Bangs and Quiet Canadians: Power, Property
Relations and Anti-Encana Sabotage in Northeastern British Columbia, October 2008–
August 2009.* MA thesis, University of British Columbia,Vancouver, BC.

Globe and Mail. 2007. Murray Klippenstein on the Ipperwash Inquiry report. May 30.

Nikiforuk, Andrew. 2009. Industrial sabotage: Under attack. *Canadian Business*, August 17.

Nikiforuk, Andrew. 2014. *Saboteurs: Wiebo Ludwig's War against Big Oil*. Vancouver, BC: Greystone.

Ohio Department of Natural Resources Division of Mineral Resources Management. 2008. *Report on the Investigation of the Natural Gas Invasion of Aquifers in Bainbridge Township of Geauga County, Ohio*. Ohio Department of Natural Resources.

Thyne, Geoffrey. 2008. *Review of Phase II Hydrogeological Study*. Garfield County, CO.

CHAPTER ELEVEN: Kafka's Law

All legal documents cited in this chapter can be accessed at *www.ernstversusencana.ca*.

Alberta Office of the Information and Privacy Commissioner. 2012. Alberta Innovates–Technology Futures. Case File Numbers F4743, F4762. Order F2012-06, March 30. At www.oipc.ab.ca/downloads/documentloader. ashx?id=3044.

Boesveld, Sarah. 2011. Report gives Harper government a failing grade for transparency. *National Post*, May 10.

Centre for Law and Democracy. 2012. *Failing to Measure Up: An Analysis of Access to Information Legislation in Canadian Jurisdictions*. Halifax, NS: Centre for Law and Democracy. At www.law-democracy.org/live/wp-content/uploads/2012/08/Canada-report-on-RTI.pdf.

Economist. 2014. Stephen Harper: The political predator. September 13.

Kleiss, Karen. 2014. Most Alberta freedom of information requests get no results, "No records exist" for two-thirds of users. *Edmonton Journal*, July 10.

McClure, Matthew. 2012. Ruling advances woman's tainted water lawsuit: Documents must be released, says Alberta information boss. *Calgary Herald*, April 17.

Talisman Energy. 2010. *Talisman Terry's Energy Adventure*. Talisman Energy Good Neighbor Program. At old.post-gazette.com/pg/pdf/201106/201106talisman_coloringbook.pdf.

CHAPTER TWELVE: The Road of the Dishes

Bachu, Stefan, and Theresa Watson. 2007. Factors affecting or indicating potential wellbore leakage. Presented to the 3rd IEA-GHG Well Bore Integrity Network Meeting, March 12–13, Santa Fe, NM.

British Columbia Oil and Gas Commission. 2012. *Investigation of Observed Seismicity in the Horn River Basin.* Victoria, BC: BC Oil and Gas Commission.

Canadian Association of Petroleum Producers. 1996. *Migration of Methane into Groundwater from Leaking Production Wells Near Lloydminster. Report for Phase 2* (1995). CAPP Pub. #1996-0003. Canadian Association of Petroleum Producers.

Cantarow, Ellen. 2013. Meet Anthony Ingraffea—From industry insider to implacable fracking opponent. EcoWatch, January 2. At ecowatch.com/2013/01/02/industry-insider-to-fracking-opponent/.

Crowe, A. S., K. A. Schaefer, A. Kohut, S. G. Shikaze, and C. J. Ptacek. 2003. *Groundwater Quality.* Linking Water Science to Policy Workshop Series. Report No. 2. Winnipeg, MB: Canadian Council of Ministers of the Environment.

Dusseault, Maurice B., Malcolm N. Gray, and Pawel A. Nowrocki. 2000. Why oilwells leak: Cement behavior and long-term consequences. Presented to the International Oil and Gas Conference and Exhibition, 7–10 November, Beijing, China. SPE-64733-MS. Society of Petroleum Engineers. doi:10.2118/64733-MS.

Dusterhoff, Dale, Greg Wilson, and Ken Newman. 2002. Field study on the use of cement pulsation to control gas migration. Presented to the SPE Gas Technology Symposium, 30 April–2 May, Calgary, AB. Society of Petroleum Engineers. doi:10.2118/75689-MS.

Dyck, Willy, and Colin E. Dunn. 1986. Helium and methane anomalies in domestic water wells in southwestern Saskatchewan, Canada, and their relationship to other dissolved constituents, oil and gas fields, and tectonic patterns. *Journal of Geophysical Research* 91, no. B12: 12343–53. doi:10.1029/JB091iB12p12343.

Ernst Environmental Services. 2013. *Brief Review of Threats to Canada's Groundwater from the Oil and Gas Industry's Methane Migration and Hydraulic Fracturing.* At www.ernstversusencana.ca/wp-content/uploads/2013/06/Brief-review-of-threats-to-Canadas-groundwater-from-oil-gas-industrys-methane-migration-and-hydraulic-fracturing-v4.pdf.

Gold, Russell. 2014. *The Boom: How Fracking Ignited the American Energy Revolution and Changed the World*. New York: Simon & Schuster.

Gruver, Mead. 2011. Wyoming air pollution worse than Los Angeles due to gas drilling. Associated Press, March 9.

Ingraffea, Anthony R. 2012. Fluid migration mechanisms due to faulty well design and/or construction: An overview and recent experiences in the Pennsylvania Marcellus Play. Physicians, Scientists & Engineers for Healthy Energy, October. At psehealthyenergy.org/site/view/1057.

Leitrim Observer. 2012. The EPA can't assess fracking in Leitrim—Jessica Ernst. March 6.

Natural Gas Europe. 2012. Tamboran Resources: The luck of the Irish. January 24. At www.naturalgaseurope.com/tamboran-resources-shale-gas-exploration-ireland.

Nikiforuk, Andrew. 2013. The shale gale is a retirement party. *The Tyee*, March 27.

Northern Ireland Assembly, Committee for Enterprise, Trade and Investment. 2012. *Official Report (Hansard)*. June 28. At www.niassembly.gov.uk/assembly-business/official-report/com-mittee-minutes-of-evidence/session-2011-2012/june-2012/shale-gas-exploration--tamboran-resources-ltd/.

Osborn, Stephen G., Avner Vengosh, Nathaniel R. Warner, and Robert B. Jackson. 2011. Methane contamination of drinking water accompanying gas-well drilling and hydraulic fracturing. *Proceedings of National Academy of Sciences* 108, no. 20 (May 9). doi:10.1073/pnas.1100682108PNAS.

Rogers, Deborah. 2013. Externalities of shales Road damage. Energy Policy Forum, April 1. At energypolicyforum.com/2013/04/01/externalities-of-shales-road-damage/.

Saskatchewan Research Council. 1995. *Migration of Methane into Groundwater from Leaking Production Wells Near Lloydminster*. CAPP Pub. #1995-0001. Canadian Association of Petroleum Producers.

Schmitz, Ron, Patrick B. Carlson, Garry D. Lorenz, Michael D. Watson, and Brian P. Erno. 1993. Husky Oil's gas migration research effort: An update. Presented to the Heavy Oil and Oil Sands Technical Symposium, March 9, Calgary, AB.

URS Operating Services, Inc. 2010. *Expanded Site Investigation—Analytical Results Report. Pavillion Area Groundwater Investigation. Pavillion, Fremont County*,

Wyoming. TDD No. 0901-01. U.S. Environmental Protection Agency Contract No. EP-W-05-050. Denver, CO: URS Operating Services.

Watson, Theresa. 2013. Alberta regulations: Wellbore integrity issues driving regulatory change. Presented to the North American Wellbore Integrity Workshop, October 16–17, Denver, CO.

Watson, Theresa, and Stefan Bachu. 2008. Wellbore leakage potential in CO_2 storage or EOR. Presented to the Fourth Wellbore Integrity Network Meeting, March 19, Paris, France.

Zuckerman, Gregory. 2013. *The Frackers: The Outrageous Inside Story of the New Billionaire Wildcatters.* New York: Penguin.

CHAPTER THIRTEEN: "No Duty of Care"

All legal documents cited in this chapter can be accessed at www.ernstversusencana.ca.

Barrett, Andrew. 2012. Comment on the National Greenhouse and Energy Reporting Measurement Amendment Determination 2012: Consultation Draft and Departmental Commentary. Geoscience Australia Submission. Reference D2012-188859.

Busetti, Seth, Kyran Mish, Peter Hennings, and Ze'ev Reches. 2012. Damage and plastic deformation of reservoir rocks: Part 2. Propagation of a hydraulic fracture. AAPG *Bulletin* 96, no 9: 1711–32. doi:10.1306/02011211011.

Carney, Matthew, and Connie Angus 2013. Gas leak! *Four Corners,* April 1. Australian Broadcasting Commission.

Daley, Mick. 2014. Wrong way to the top. MickDaley.Com, April 13. At www.mickdaley.com/?p=148.

Dingle, Sarah. 2013. Questions raised over environmental assessments of billion dollar CSG developments. ABC News, April 1. At www.abc.net.au/am/content/2013/s3727070.htm.

Hannam, Peter. 2014. CSG study finds elevated methane levels near gas fields. *Sydney Morning Herald,* November 20. At www.smh.com.au/business/carbon-economy/csg-study-finds-elevated-methane-levels-near-gas-fields-20141120-11pj90.html.

Maher, Damien T., Isaac R. Santos, and Douglas R. Tait. 2014. Mapping methane and carbon dioxide concentrations and ẟ13C values in the atmosphere of two Australian coal seam gas fields. *Water, Air, & Soil Pollution* 225: 2216. doi:10.1007/s11270-014-2216-2.

Maher, Damien, Douglas Tait, and Isaac Santos. 2014. Science and coal seam gas—a case of the tortoise and the hare? *The Conversation*, December 7. At theconversation.com/ science-and-coal-seam-gas-a-case-of-the-tortoise-and-the-hare-35100.

Michelin, Lana. 2011. Couple claims energy exploration contaminates their water. *Red Deer Advocate*, December 26.

New South Wales Chief Scientist & Engineer. 2013. *Initial Report on the Independent Review of Coal Seam Gas Activities in* NSW. Sydney, Australia: NSW Chief Scientist & Engineer.

Office of the Ethics Commissioner, Province of Alberta. 2013. *Report to the Speaker of the Legislative Assembly of Alberta of the Investigation by Neil Wilkinson, Ethics Commissioner into Allegations Involving the Honourable Alison Redford, Q.C., Premier.* December 4. At www.ethicscommissioner.ab.ca/media/1063/final-ver-01-dec-03-13.pdf.

Protti, Gerard. 2005. Letter to the editor. *Report on Business Magazine*, June 24.

Schneiders, Lyndon. 2014. Quest for coal seam gas sullies name of Santos. *Sydney Morning Herald*, March 17.

Tait, Douglas R., Isaac R. Santos, Damien T. Maher, Tyler J. Cyronak, and Rachael J. Davis. 2013. Enrichment of radon and carbon dioxide in the open atmosphere of an Australian coal seam gas field. *Environmental Science and Technology* 47, no. 7: 3099–3104. doi:10.1021/es304538g.

Yedlin, Deborah. 2013. Chair of Alberta's new energy regulator a real industry insider. *Calgary Herald*, April 2.

CHAPTER FOURTEEN: The Sisters of Jessica Ernst

Alberta. 2013. *Alberta Oil and Gas Industry: Quarterly Update.* Spring. Calgary, AB.

Colburn, Theo, Carol Kwiatkowski, Kim Schultz, and Mary Bachran. 2011. Natural gas operations from a public health perspective. *Human and Ecological Risk Assessment* 17, no. 5: 1039–56. doi:10.1080/10807039.2011.605662.

Court of Queen's Bench of Alberta. 2013. Statement of Claim, Diana Daunheimer vs. Angle Energy Incorporated, December 6. Court File Number 1301-14429.

Daunheimer, Diana. 2013. Environmentally accountable Fracking a must [letter to the editor]. *Mountain View Gazette*, November 12.

Daunheimer, Diana. 2014. Hydraulic fracturing and how it affects us all. Presented to the Alberta Surface Rights Federation, March 13, Camrose, AB.

Gleeson, John. 2011. Eagle Valley woman Kim Mildenstein urges action on traffic concerns. *Mountain View Gazette*, November 8.

Gleeson, John. 2012. Ernst learned of alleged threat prior to Eagle Hill appearance. *Mountain View Gazette*, March 27.

Hosegood, Margaret. 2013. Kudos for Alberta's energy workers [letter to the editor]. *Mountain View Gazette*, July 9.

Kassotis, Christopher D., Donald E. Tillitt, J. Wade Davis, Annette M. Hormann, and Susan C. Nagel. 2013. Estrogen and androgen receptor activities of hydraulic fracturing chemicals and surface and ground water in a drilling-dense region. *Endocrinology*, December 16. doi:10.1210/en.2013-1697.

Macey, Gregg P., Ruth Breech, Mark Cherniak, et al. 2014. Air concentrations of volatile compounds near oil and gas production: A community-based exploratory study. *Environmental Health* 13: 82. doi:10.1186/1476-069x-13-82.

National Energy Board. 2011. Energy briefing note: Tight oil developments in western Canadian sedimentary basin. December.

Nikiforuk, Andrew. 2014. Alberta mother fights five neighbouring fracked wells. *The Tyee*, February 28.

Perry, Simona. 2011. Fractured communities: Fractured lives. Presented to the Summit on the Impacts of Fracking, 27th Annual Meeting of Clean Water for NC, September 10. At vimeo.com/30081811.

Provincial Court of Alberta Judicial Centre of Calgary. 2012. Her Majesty the Queen v. Kimberly Joan Mildenstein, Didsbury, Alberta. *Proceedings*, October 22.

Vink, Kevin. 2013. County resident takes on Angle. *Mountain View Gazette*, July 9.

EPILOGUE: Completions

Chu, Jennifer. 2014. An extinction in the blink of an eye. MIT News Office, February 10.

Court of Queen's Bench of Alberta. 2014. Ernst v. EnCana Corporation. ABQB 672.

Drajem, Mark, and Jim Efstathiou. 2013. Drillers silence fracking claims with sealed settlements. *Bloomberg Business*, June 5.

Ellul, Jacques. 1990. *The Technological Bluff*. Translated by Geoffrey W. Bromiley. Grand Rapids, MI: Eerdmans.

Gass, Henry. 2014. How scientists overlooked a 2,500 square mile cloud of methane over the Southwest. *Christian Science Monitor*, October 10.

Hannam, Peter. 2015. Senators move to give CSG whistleblower air. *Sydney Morning Herald*, March 2.

Howarth, Robert, Renee Santoro, and Anthony Ingraffea. 2011. Methane and the greenhouse-gas footprint of natural gas from shale formations: A letter. *Climatic Change* 106: 679–90. doi:10.1007/s10584-011-0061-5.

Jackson, Robert, and Dominic DiGiulio. 2014. Hydraulic fracturing in underground sources of drinking water at Pavillion, Wyoming. Presented to the American Chemical Society National Meeting and Exposition, Evolving Science and Environmental Impact of Hydraulic Fracturing, August 10–14, San Francisco, California.

Keranen, K. M., M. Weingarten, G. A. Abers, B. A. Bekins, and S. Ge. 2014. Sharp increase in central Oklahoma seismicity since 2008 induced by massive wastewater injection. *Science* 345, no. 6195: 448–51. doi:10.1126/science.1255802.

Kirschke, Stefanie, Philippe Bouquet, Philippe Ciais, et al. 2013. Three decades of global methane sources and sinks. *Nature Geoscience* 6: 813–23. doi:10.1038/ngeo1955.

Kort, Eric A., Christian Frankenberg, Keeley R. Costigan, et al. 2014. Four Corners: The largest US methane anomaly viewed from space. *Geophysical Research Letters*, 9 October. doi:10.1002/2014GL061503.

Meng, Yanjun, Dazhen Tang, Hao Xu, Yong Li, and Lijun Gao. 2014. Coalbed methane produced water in China: Status and environmental issues. *Environmental Science and Pollution Research* 21: 6964–74. doi:10.1007/s11356-014-2675-4.

Moore, Robert, Ian Palmer, and Nigel Higgs. 2014. Anisotropic model for permeability change in coalbed methane wells. Presented to the SPE Western North American and Rocky Mountain Joint Meeting, 17–18 April, Denver, Colorado. SPE-169592-MS. Society of Petroleum Engineers. doi:10.2118/169592-MS.

Moore, Tony. 2014. Whistleblower says there was evidence of CSG breaches. *Brisbane Times*, June 27.

Nikiforuk, Andrew. 2014. Ailing shale gas returns force a drilling treadmill. *The Tyee*, January 27.

Nikiforuk, Andrew. 2015. Did Alberta just break a fracking earthquake world record? *The Tyee*, January 29.

Polczer, Shaun. 2012. Canada's long forgotten coal-bed methane play. *Petroleum Economist*, July/August.

Preece, Rob. 2012. The Door to Hell. *Daily Mail*, July 27.

Ridlington, Elizabeth, and John Rumpler. 2013. *Fracking by the Numbers: Key Impacts of Dirty Drilling at the State and National Level*. Boston: Environment America Research & Policy Center. At www.environmentamerica.org/sites/environment/files/reports/EA_FrackingNumbers_scrn.pdf.

Sayles, Robin. 2013. French coalbed gas drilling presents risks even without fracking: land agencies. Platts, November 11. At www.platts.com/latest-news/natural-gas/london/french-coalbed-gas-drilling-presents-risks-even-27619068.

Schlesinger, Joel. 2014. Process raises concerns over water usage, quality. http://www.csur.com/images/news/CAPP_CH_Special%20Report_Aug%2028%202014.pdf.

INDEX

AccuMap, 137

acid stimulation, 38–39

adrenal cancer, 120

AER, see Alberta Energy Regulator

Ahmed, Usman, 307

air pollution, 111, 234, 281–82

Alabama, see Black Warrior Basin

Alberta: call for stricter regulations in, 257–58; Cardium Shale, 279–87; confidentiality agreements in, 303; Duvernay Formation, 304; earthquakes in, 75; water contamination in, 179–80. See also Alberta government; Horseshoe Canyon Formation

Alberta Association of Municipal Districts and Counties, 257–58

Alberta Energy Company, 95, 124. See also Encana

Alberta Energy Regulator (AER): and Daunheimer, 283–84, 287; name changes, 315; powers given to, 261; and Protti, 174, 260. See also Energy and Utilities Board; Energy Resources Conservation Board

Alberta Environment: and ARC, 182, 184; and ARC report, 216, 232, 271; Campbell's well, 149; and CBM, 152–53; changes in name, 315; and Daunheimer, 283, 285–86; Ernst's well testing, 143–45; Jack case, 159–60; and Quicksilver Resources, 263–64; and Rosebud wells, 216; Signer's well testing, 153, 155, 157–59; and water contamination, 138, 150, 151, 162, 182, 189. See also lawsuit, Ernst

Alberta Environment and Sustainable Resource Development, see Alberta Environment

82, 83–84; McMillian lawsuit,
85–86; new regulations for,
85–86; water contamination in,
80–81
Blancett, Tweeti, 146, 147
BLM (Bureau of Land Management),
56–57, 72–73, 76, 77
Blyth, Alexander, 183, 185
Booth, John Wilkes, 33
Borden Ladner Gervais (law firm),
302
Bourget, Darren, 143–44
Boutilier, Guy, 143, 147, 161, 163
Boyle, Neil, 271
Bracken, Lisa, 193
Bradford County (PA), 298–300
Bray, Shirley, 253
Britain, 304–5
British Columbia, 75, 207–8, 234
Brooymans, Hanneke, 127–28
Bullach, Dave, 112
Bureau of Land Management (BLM),
56–57, 72–73, 76, 77
Burren, Ireland, 247
Bush administration, 86, 90
butane, 180
bystander culture, 296–97

California, 39. See also Los Angeles
Campbell, Ben Nighthorse, 75
Campbell, Ronalie, 149, 303
Campbell, Shawn: and ARC report,
185, 189, 192; support for Ernst,
249, 273, 303; water contamination, 149–50, 189

Canadian Association of Petroleum
Producers, 151, 238–39, 304
Canadian Council of Ministers of
the Environment, 162–63, 239
Canadian government, 224–25, 232,
257, 289
Canadian military, 193–94
Canadian Parks and Wilderness
Society, 125
Canadian Society for Unconven-
tional Gas, see Canadian Society
for Unconventional Resources
Canadian Society for Unconven-
tional Resources, 99, 161
cancer, adrenal, 120
carbon dioxide, 63, 183–84
carbon monoxide, 63
Cardium Shale, 279–87
cavitation, 72
CBM, see coalbed methane
cement, 39, 132, 244–45
Chafin, Daniel T., 75–77
Charter of Rights and Freedoms, 197,
225, 229–30
Cheney, Dick, 86
Cherry, John, 275
Chesapeake Energy, 233
Chilingarian, George, 6, 10, 11–12
China, 308
CIBC, 104–5
Class II injection wells, 79, 83. See
also fluid injection
Clissold, Roger, 163
CO2 fracking, 183–84
coal, 59–61, 68, 188

337

339

DAVID SUZUKI INSTITUTE

The David Suzuki Institute is a non-profit organization founded in 2010 to stimulate debate and action on environmental issues. The Institute and the David Suzuki Foundation both work to advance awareness of environmental issues important to all Canadians.

We invite you to support the activities of the Institute. For more information please contact us at:

David Suzuki Institute
219 – 2211 West 4th Avenue
Vancouver, BC, Canada v6k 4s2
info@davidsuzukiinstitute.org
604-742-2899
www.davidsuzukiinstitute.org

Cheques can be made payable to The David Suzuki Institute.